Iason Papaioannou

Non-intrusive finite element reliability analysis

Iason Papaioannou

Non-intrusive finite element reliability analysis

Structural reliability analysis with "black box" finite element programs

Südwestdeutscher Verlag für Hochschulschriften

Impressum / Imprint
Bibliografische Information der Deutschen Nationalbibliothek: Die Deutsche Nationalbibliothek verzeichnet diese Publikation in der Deutschen Nationalbibliografie; detaillierte bibliografische Daten sind im Internet über http://dnb.d-nb.de abrufbar.
Alle in diesem Buch genannten Marken und Produktnamen unterliegen warenzeichen-, marken- oder patentrechtlichem Schutz bzw. sind Warenzeichen oder eingetragene Warenzeichen der jeweiligen Inhaber. Die Wiedergabe von Marken, Produktnamen, Gebrauchsnamen, Handelsnamen, Warenbezeichnungen u.s.w. in diesem Werk berechtigt auch ohne besondere Kennzeichnung nicht zu der Annahme, dass solche Namen im Sinne der Warenzeichen- und Markenschutzgesetzgebung als frei zu betrachten wären und daher von jedermann benutzt werden dürften.

Bibliographic information published by the Deutsche Nationalbibliothek: The Deutsche Nationalbibliothek lists this publication in the Deutsche Nationalbibliografie; detailed bibliographic data are available in the Internet at http://dnb.d-nb.de.
Any brand names and product names mentioned in this book are subject to trademark, brand or patent protection and are trademarks or registered trademarks of their respective holders. The use of brand names, product names, common names, trade names, product descriptions etc. even without a particular marking in this works is in no way to be construed to mean that such names may be regarded as unrestricted in respect of trademark and brand protection legislation and could thus be used by anyone.

Coverbild / Cover image: www.ingimage.com

Verlag / Publisher:
Südwestdeutscher Verlag für Hochschulschriften
ist ein Imprint der / is a trademark of
AV Akademikerverlag GmbH & Co. KG
Heinrich-Böcking-Str. 6-8, 66121 Saarbrücken, Deutschland / Germany
Email: info@svh-verlag.de

Herstellung: siehe letzte Seite /
Printed at: see last page
ISBN: 978-3-8381-3541-0

Zugl. / Approved by: München, TU, Diss., 2012

Copyright © 2013 AV Akademikerverlag GmbH & Co. KG
Alle Rechte vorbehalten. / All rights reserved. Saarbrücken 2013

To the memory of my father
Ioannis Papaioannou

Preface

In order to motivate the incorporation of the reliability concept in the structural design and analysis procedures, the engineering community is in need of finite element (FE) software with capabilities of including the stochastic nature of input parameters. This work presents a series of FE reliability methods that have been implemented in a reliability tool and integrated into a commerical FE software package. The tool is programmed in a stand-alone fashion and utilizes the FE solver as "black box". Hence, the applied reliability methods are termed non-intrusive, since they do not have access to the core routines of the FE software.

The presentation starts with a review of probability theory. The modeling of spatial variability using random fields is then addressed and a series of existing random field discretization methods are evaluated. Moreover, the possibility of an embedded-domain discretization of random fields is examined. Furthermore, the basic concepts of reliability theory are introduced, followed by a detailed presentation of the implemented reliability methods. These include the FORM/SORM, combined with robust optimization algorithms, as well as a variety of simulation techniques, including directional simulation, importance sampling and subset simulation. Next, the problem of updating the reliability estimate conditional on measurement or other information is discussed. Finally, the methods are applied for the reliability assessment of a number of nonlinear geotechnical FE models.

Preface viii

Acknowledgements

This monograph is based on a PhD thesis conducted at the Chair of Computation in Engineering of the Technische Universität München.

First of all, I would like to thank Prof. Dr.rer.nat. Ernst Rank for offering me the opportunity to work on such a fascinating topic as well as for his continuous support throughout my MSc and PhD years.

I would like to thank Prof. Dr.sc.techn. Daniel Straub, whose guidance and enthusiasm inspire my working days. I would also like to thank him for agreeing to be the second examiner of my PhD thesis.

Special thanks to Prof. Manolis Papadrakakis who agreed to act as an examiner of the thesis.

Also, I would like to express my deepest thanks to Prof. Dr.-Ing. Casimir Katz and my supervisor from SOFiSTiK Dr.-Ing. Holger Heidkamp, whose interest on the topic led to the financing of my PhD. Moreover, the numerous discussions with Holger Heidkamp and his ongoing interest on the subject were invaluable.

I would like to thank Prof. Armen Der Kiureghian for his guidance and support during my visit at the Deparment of Civil Engineering of the UC Berkeley. The months I spent in Berkeley were very influential in my knowledge on probabilistic engineering. I would like to thank Prof. Rank and Prof. Straub, who both made this visit possible.

Thanks Christian for being a friend and the best officemate. Thanks Spyros for the tolerance and support.

Thanks to all my friend and colleagues for their support and encouragement: André Borrmann, Hanne Cornils, Geli Kneidl, Eleni Palkopoulou, Maro Papageorgiou, Patty Papakosta.

A few years ago, while studying my first degree at the Technical University of Athens, and quite frankly feeling disappointed in the perspective of pursuing a career as a civil engineer, I discovered a passion for scientific research. Back then, there were a few people that urged me to follow my newly found passion and apply for a research-oriented master course at the Technische Universität München. These include Prof. Manolis Papadrakakis and Dr. Michalis Fragiadakis, who co-supervised by diploma thesis in Athens and initiated me to the world of scientific research. I would like to thank them both for guiding me through my first scientific steps.

Finally, I would like to thank my family for their love and support; my sister Margarita, my mother Marina and my late father Ioannis.

Iason Papaioannou
January, 2013

Contents

Preface vii

Acknowledgements ix

Contents xi

1   **Introduction** 1
  1.1   Motivation and scope 1
  1.2   Outline 4

2   **Modeling of uncertainties** 7
  2.1   Basic notions of probability 7
  2.2   Random variables 9
      2.2.1   Basic definitions 9
      2.2.2   Commonly used distributions 12
  2.3   Random vectors 20
      2.3.1   Joint distribution and joint moments 20
      2.3.2   Transformation of random vectors 23
      2.3.3   The multinormal distribution 24
      2.3.4   The Nataf model 26
  2.4   Random processes and random fields 27
      2.4.1   Basic definitions 27
      2.4.2   Homogeneous random fields 29
      2.4.3   Derivatives of random fields 34

    2.4.4 Integrals of random fields 38
    2.4.5 Gaussian random fields 39
    2.4.6 Non-Gaussian random fields 41
2.5 Discretization of random fields 42
    2.5.1 Point discretization methods 43
    2.5.2 Average discretization methods 47
    2.5.3 Series expansion methods 50
    2.5.4 Comparison of the discretization methods 57
    2.5.5 Embedded-domain discretization of random fields 62

3 Fundamental concepts of reliability analysis 71
3.1 Evolution of design concepts 71
3.2 The elementary reliability problem 74
3.3 The generalized reliability problem 76
    3.3.1 Generalization of the probability of failure 78
    3.3.2 Reliability measures 79
3.4 The system reliability problem 80

4 Finite element reliability assessment 83
4.1 Introductory comments 83
4.2 Isoprobabilistic transformation 85
4.3 The first order reliability method 87
    4.3.1 Optimization algorithms 91
    4.3.2 Comparison of the optimization algorithms 97
    4.3.3 Sensitivity measures 99
4.4 The inverse first order reliability method 101
4.5 The second order reliability method 103
4.6 System reliability using FORM 105
4.7 Simulation methods 107
    4.7.1 Generation of random samples 108
    4.7.2 The Monte Carlo method 110
    4.7.3 Directional simulation 114
    4.7.4 Importance sampling methods 118
    4.7.5 Comparison of the simulation methods 131
4.8 Simulation in high dimensions 136

xiii   Contents

    4.8.1 The subset simulation 137
4.9   Response surface methods 141

5   Bayesian updating of reliability estimates 147
5.1   Introduction 147
5.2   Updating with equality information 148
    5.2.1 First- and second-order methods 148
    5.2.2 A general approach 150

6   Numerical examples 153
6.1   Reliability analysis of a deep circular tunnel 153
    6.1.1 Finite element solution of tunnel deformation analysis 154
    6.1.2 Results without spatial variability considerations 155
    6.1.3 Finite element model including spatial variability 157
    6.1.4 Results accounting for spatial variability of soil 159
6.2   Reliability updating of a cantilever embedded wall 161
    6.2.1 Finite element model 162
    6.2.2 Limit-state functions 165
    6.2.3 Results 166
6.3   Reliability-based design of slope angle 169
    6.3.1 Finite element slope stability analysis 170
    6.3.2 Results 173

7   Conclusion 179

Bibliography 183

1 Introduction

1.1 Motivation and scope

In structural engineering analysis and design, the goal is to predict the behavior of the structural system subject to certain prescribed loading conditions. This task requires the mathematical modeling of a physical phenomenon, which consists of a set of partial differential equations whose solution aims at reproducing experimental observations. In complex structural systems, a numerical solution of the mathematical model is usually sought by application of finite element (FE) methods. The increase of the computational power has motivated the development of FE methods that are able to capture the behavior of models of ever increasing complexity. These methods have been implemented in commercial software that are widely used in the structural engineering community.

However, despite the increase in the accuracy of the models and the respective FE representations, they will never be able to capture the complexity of the real world. This is due to the fact that neither the loading nor some structural parameters can be known with certainty, i.e. they are not deterministic quantities. Therefore, the structural response inherits this property and hence should be considered an uncertain quantity. This implies that FE solutions at best provide an approximation of the actual reality.

In civil engineering structures, the relevant quantities that cannot be known with certainty are usually of either one of the following three kinds:

- Material properties; a certain deviation between the material properties of a structure and the standardized properties of the material are expected.
- Loads; idealization of the applied loads will never capture the complexity of the actual loading conditions.
- Geometric dimensions; imperfections in the construction phase will lead to deviations between the geometry of the design and the actual structure.

The uncertainties that effect these parameters can be categorized into two distinct types; namely, aleatory and epistemic uncertainties. Aleatory uncertainties refer to the intrinsic randomness of a phenomenon. As an example, submitting concrete specimens of the same batch to the same material test under the same experimental conditions will lead to different estimates of its strength. Aleatory uncertainties cannot be reduced by collecting additional information. On the other hand, epistemic uncertainties refer to a lack of knowledge. In the aforementioned example, the uncertainty in the prediction of the concrete strength can be reduced by increasing the accuracy of the measurement device or by performing a larger number of tests. That is, epistemic uncertainty can be reduced by collecting additional data or by refining the models. In many cases, it is difficult to distinguish between these two different types of uncertainties that may be present in the same structural system. A discussion on the modeling of these different types of uncertainty is given in [23]. Within this study, such a distinction is not made, although in principle the methods presented here can be applied to the case where both types are present.

Structural engineers usually account for the uncertainties in the loading and structural properties by applying safety factors, which are chosen following the directives of published design codes. This approach is motivated by its computational efficiency, since it involves a single FE calculation. However, safety factors do not allow a quantification of the uncertainty of response quantities and they can lead to over-conservative designs. A more rigorous approach involves the explicit consideration of the relevant uncertainties and attempts a complete characterization of uncertain response quantities. This approach can be realized with the help of notions of probability theory.

In most cases, the relevant information sought is the probability that a response quantity exceeds a prescribed threshold. The threshold is chosen by the engineer to ensure that its non-exceedance will correspond to a successful performance of the structure. The aforementioned probability is termed probability of failure and its complementary probability, i.e. the probability of successful performance, is referred to as reliability. The process of computing the reliability of a structural system is termed reliability analysis ([35], [71], [83]).

Reliability analysis provides a sound basis for quantifying the risk of civil engineering structures and usually leads to reduced design costs, as compared to designs based on conventional safety factors. However, the evaluation of the reliability or the probability of failure is usually not a straightforward task, especially when it involves complex engineering structures that are modeled by FE methods. Solution methods can be classified into two categories; intrusive and non-intrusive. Intrusive methods involve reprogramming of the FE software while non-intrusive methods utilize the FE software as "black boxes". The latter methods have the advantage that they can be combined with existing and widely-accepted FE software. However, they may involve high computational costs, since for complex engineering structures the single deterministic FE calculation is usually time-consuming.

This work focuses on non-intrusive methods for FE reliability analysis. Non-intrusive FE reliability methods can be divided into two categories. The first involves first- or second-order approximations of the failure condition of the structure. Methods belonging to this category require the solution of an optimization problem and provide an approximation of the actual reliability [20]. The main computational cost of these methods is related to the evaluation of the derivatives at each step of the optimization algorithm. The second category is based on the repeated generation of possible outcomes of the uncertain variables [112]. For each outcome, the FE solver is called and the failure condition is evaluated. These methods are generally applicable but in most cases involve considerable computational cost.

The methods presented herein have been implemented in a FORTRAN program [96] that is fully integrated to the commercial FE software package SOFiSTiK [126]. The program involves a general purpose implementation of non-intrusive reliability methods through a "black box" coupling to the FE solver.

1.2 Outline

This book is organized into seven chapters, which are outlined as follows:

Chapter 2 first provides an introduction to the basics of the theory of probability and random variables and reviews basic distribution models. Then, the representation of quantities that vary randomly in space or time is addressed by introducing the mathematical notion of random fields. Furthermore, methods for the numerical treatment of random fields are reviewed and assessed through numerical studies. Finally, a novel idea for the treatment of random fields in non-standard domains is investigated.

Chapter 3 presents the fundamental concepts of reliability analysis. First, a review of the existing design concepts is provided, followed by a detailed presentation of the component and system reliability problems.

Chapter 4 describes the implemented FE reliability analysis methods. Approximation methods are presented first, with a focus on the algorithms for the solution of the underlying optimization problem. Subsequently, a detailed review of simulation methods is presented. The special case of simulation in problems with a large number of random variables is treated next. Finally a review of response surface methods with application to reliability analysis is provided.

Chapter 5 introduces methods for updating the reliability of a structural system, in light of new information. First, the reliability updating problem is formulated. Next a reliability updating method based on approximation concepts is presented, followed by a general approach that can be treated with any of the methods described in Chapter 4.

Chapter 6 presents three industrial applications, which demonstrate the potential of the methods described in the previous chapters. The first consists of the reliability analysis of a deep tunnel in rocky soil. The second introduces a framework for reliability updating of geotechnical construction sites with application to a deep excavation with a cantilever retaining wall. The third example proposes a method for the reliability-based design of slope angle, including the spatial variability of the soil material.

Finally, Chapter 7 presents a summary of the contributions of this work.

2 Modeling of uncertainties

The first step in the reliability assessment of structural systems is the probabilistic description of the input parameters that are expected to present an uncertain behavior. This chapter contains the mathematical notions and the different procedures that are utilized for the modeling of the uncertain parameters. Although some basics of probability theory are provided, these are limited to the minimal requirements of the present document. For a more detailed view in probability theory, the reader is referred to standard texts (e.g. [100]).

2.1 Basic notions of probability

Consider a sample space S, containing all possible outcomes of an experiment. An event E is defined as any subset of S (i.e. $E \subseteq S$). The basic operations on two events E_1, E_2 are the union $E_1 \cup E_2$ and the intersection $E_1 \cap E_2$. These operations have the following distributive properties:

$$E_1 \cap (E_2 \cup E_3) = (E_1 \cap E_2) \cup (E_1 \cap E_3)$$
$$E_1 \cup (E_2 \cap E_3) = (E_1 \cup E_2) \cap (E_1 \cup E_3) \tag{2.1}$$

2 Modeling of uncertainties

The events $E_1, \ldots, E_n$ are called mutually exclusive if:

$$E_i \cap E_j = \emptyset \qquad \forall i,j : 1 \leq i < j \leq n \qquad (2.2)$$

where $\emptyset$ is the empty set. The events $E_1, \ldots, E_n$ are called collectively exhaustive if:

$$\bigcup_{i=1}^{n} E_i = S \qquad (2.3)$$

The set of events containing all individual outcomes of the experiment is a complete set of mutually exclusive and collectively exhaustive events. The same holds for any event E and its complementary event $\overline{E}$, i.e.:

$$\begin{aligned} E \cap \overline{E} &= \emptyset \\ E \cup \overline{E} &= S \end{aligned} \qquad (2.4)$$

The following rules hold for the complement of union and intersection of events:

$$\begin{aligned} \overline{\bigcup E_i} &= \bigcap \overline{E_i} \\ \overline{\bigcap E_i} &= \bigcup \overline{E_i} \end{aligned} \qquad (2.5)$$

The set of all events in the sample space S defines a σ-algebra F. The probability space associated with S is a measure space, consisting of the triple (S, F, P), where $P : F \rightarrow [0,1]$ is the probability measure, defined as a measure of the likelihood of occurrence of any event $E \in F$. The measure P follows the Kolmogorov axioms:

$$P(E) \geq 0 \qquad \forall E \in F \qquad (2.6)$$

$$P(S) = 1 \qquad (2.7)$$

$$P\left(\bigcup_i E_i\right) = \sum_i P(E_i) \qquad \forall E_1, E_2, \ldots \in F : \{E_1, E_2, \ldots\} \text{ mut. excl.} \qquad (2.8)$$

From the Kolmogorov axioms, the following important results follow:

$$P(\bar{E}) = 1 - P(E) \tag{2.9}$$

$$P(\emptyset) = 0 \tag{2.10}$$

$$P(E_1 \cup E_2) = P(E_1) + P(E_2) - P(E_1 \cap E_2) \quad \forall E_1, E_2 \in F \tag{2.11}$$

The conditional probability of an event E_1 given that the event E_2 has occurred is defined by:

$$P(E_1 | E_2) = \frac{P(E_1 \cap E_2)}{P(E_2)} \tag{2.12}$$

Reformulation of Eq. (2.12) leads to the following multiplication rule:

$$\begin{aligned} P(E_1 \cap E_2) &= P(E_1 | E_2) P(E_2) \\ &= P(E_2 | E_1) P(E_1) \end{aligned} \tag{2.13}$$

We can generalize Eq. (2.13) for the events $E_1, \ldots, E_n$ as follows:

$$\begin{aligned} P(E_1 \cap \ldots \cap E_n) &= P(E_1 | E_2 \cap \ldots \cap E_n) \ldots P(E_{n-1} | E_n) P(E_n) \\ &= P(E_n | E_1 \cap \ldots \cap E_{n-1}) \ldots P(E_2 | E_1) P(E_1) \end{aligned} \tag{2.14}$$

Two events E_1, E_2 are said to be statistically independent if:

$$P(E_1 | E_2) = P(E_1) \tag{2.15}$$

A direct consequence of Eq. (2.15) is the following:

$$P(E_1 \cap E_2) = P(E_1) P(E_2) \tag{2.16}$$

2.2 Random variables

2.2.1 Basic definitions

A random variable X is defined as a function that maps elements of a sample space S (i.e. individual outcomes of the experiment) to the set of real numbers, i.e. $X : \text{S} \to \mathbb{R}$. The space S, and correspondingly the random variable X, can either be discrete or continuous. Denoting an individual outcome of S by x, we can define the following event:

$$E = \{X \leq x\} \tag{2.17}$$

The function that expresses the probability $P(E)$ of occurrence of this event in terms of the outcome x is the cumulative distribution function (CDF) $F_X(x)$ of the random variable X, i.e.

$$F_X(x) = P(X \leq x) \tag{2.18}$$

A random variable can be completely defined by its CDF. The CDF is a non-decreasing function with the following properties:

$$\lim_{x \to -\infty} F_X(x) = 0 \quad \text{and} \quad \lim_{x \to +\infty} F_X(x) = 1 \tag{2.19}$$

For discrete random variables, we can define the probability mass function (PMF) as follows:

$$p_X(x) = P(X = x) \tag{2.20}$$

wherein the probability $P(X = x)$ has a finite value due to the discrete nature of the corresponding sample space. For the PMF, the following normalization rule holds:

$$\sum_{\forall i} p_X(x_i) = 1 \tag{2.21}$$

For a discrete random variable, the CDF is defined as follows:

$$F_X(x) = \sum_{\forall x_i \leq x} p_X(x_i) \tag{2.22}$$

In continuous sample spaces, probabilities of the type $P(X = x)$ are zero. Therefore, for a continuous random variable, the probability density function (PDF) is defined as follows:

$$f_X(x) = \lim_{dx \to 0} \frac{P(x < X \leq x + dx)}{dx} \tag{2.23}$$

2.2 Random variables

Hence, the PDF can be obtained by differentiation of the CDF:

$$f_X(x) = \frac{dF_X(x)}{dx} \tag{2.24}$$

The CDF can therefore be defined in this case as:

$$F_X(x) = \int_{-\infty}^{x} f_X(z)\,dz \tag{2.25}$$

Note that, as shown in Eq. (2.23)–(2.25), the integral of the PDF and not the function itself expresses a meaningful probability measure. For the PDF, the normalization rule reads:

$$\int_{-\infty}^{\infty} f_X(x)\,dx = 1 \tag{2.26}$$

Physical quantities are usually associated with continuous sample spaces, therefore in this text we will mostly deal with continuous random variables.

Let $g(X)$ be any function of the random variable X. The mathematical expectation $E[g(X)]$ is defined for a discrete random variable as:

$$E[g(X)] = \sum_{\forall i} g(x_i) p_X(x_i) \tag{2.27}$$

and for a continuous random variable as:

$$E[g(X)] = \int_{-\infty}^{\infty} g(x) f_X(x)\,dx \tag{2.28}$$

The mean (expected) value of a random variable can then be defined as:

$$\mu_X = E[X] \tag{2.29}$$

The n-th moment μ_n and n-th central moment μ'_n of X are defined as follows:

$$\mu_n = E[X^n] \tag{2.30}$$

$$\mu'_n = \mathrm{E}\left[(X-\mu_X)^n\right] \qquad (2.31)$$

We can then define the variance Var(X) and standard deviation σ_X of X as follows:

$$\mathrm{Var}(X) = \mathrm{E}\left[(X-\mu_X)^2\right] \qquad (2.32)$$

$$\sigma_X = \sqrt{\mathrm{Var}(X)} \qquad (2.33)$$

The variance is a measure of the dispersion of a PDF. For random variables with non-zero mean value ($\mu_X \neq 0$), we can define the dimensionless coefficient of variation CV_X as:

$$CV_X = \frac{\sigma_X}{|\mu_X|} \qquad (2.34)$$

The normalized third central moment of X is the skewness coefficients γ_X, defined as follows:

$$\gamma_X = \frac{\mathrm{E}\left[(X-\mu_X)^3\right]}{\sigma_X^3} \qquad (2.35)$$

If a random variable has $\gamma_X = 0$, then its PDF is symmetric about the mean. If $\gamma_X < 0$, then the left tail is longer, while if $\gamma_X > 0$ the right tail is longer.

The normalized fourth central moment is the kurtosis coefficient κ_X, defined as follows:

$$\kappa_X = \frac{\mathrm{E}\left[(X-\mu_X)^4\right]}{\sigma_X^4} \qquad (2.36)$$

The kurtosis coefficient is a measure of the flatness of a PDF.

2.2.2 Commonly used distributions

This section discusses a few commonly used distributions of continuous random variables.

Normal or Gaussian distribution

The Gaussian distribution is of great importance in the fields of probability and statistics. Moreover, it is one of the most frequently used distributions in engineering problems. A Gaussian random variable $X \sim N(\mu, \sigma)$ is defined by two parameters, the mean μ and the standard deviation σ. Its PDF is as follows:

$$f_X(x) = \frac{1}{\sigma}\varphi\left(\frac{x-\mu}{\sigma}\right) = \frac{1}{\sigma\sqrt{2\pi}}\exp\left[-\frac{(x-\mu)^2}{2\sigma^2}\right] \quad x \in (-\infty, +\infty) \quad (2.37)$$

where $\varphi(.)$ denotes the standard normal PDF associated with the standard normal random variable $U \sim N(0, 1)$:

$$\varphi(u) = \frac{1}{\sqrt{2\pi}}\exp\left[-\frac{u^2}{2}\right] \quad u \in (-\infty, +\infty) \quad (2.38)$$

The CDF of $X \sim N(\mu, \sigma)$ is as follows:

$$F_X(x) = \Phi\left(\frac{x-\mu}{\sigma}\right) \quad (2.39)$$

where $\Phi(.)$ denotes the standard normal CDF:

$$\Phi(u) = \int_{-\infty}^{u} \frac{1}{\sqrt{2\pi}}\exp\left[-\frac{z^2}{2}\right]dz \quad (2.40)$$

The skewness coefficient of a Gaussian random variable is $\gamma = 0$, which implies a symmetric PDF, while its kurtosis coefficient is $\kappa = 3$. These two values usually serve as measures on whether experimental data sets can be modeled by the Gaussian distribution. Quantitative representations of the normal PDF and CDF are shown in Figure 2.1.

Lognormal distribution

A lognormal random variable $X \sim LN(\lambda, \zeta)$ is a two-parameter random variable, with the following PDF and CDF, respectively:

$$f_X(x) = \frac{1}{\zeta x}\varphi\left(\frac{\ln x - \lambda}{\zeta}\right) \quad x \in (0, +\infty) \tag{2.41}$$

$$F_X(x) = \Phi\left(\frac{\ln x - \lambda}{\zeta}\right) \tag{2.42}$$

Hence, the natural logarithm of X is a Gaussian random variable, i.e. $\ln X \sim N(\lambda, \zeta)$. The mean and standard deviation of X are expressed in terms of the parameters λ, ζ as:

$$\begin{aligned} \mu &= \exp\left(\lambda + \frac{\zeta^2}{2}\right) \\ \sigma &= \mu\sqrt{\exp(\zeta^2) - 1} \end{aligned} \tag{2.43}$$

Conversely, the parameters λ, ζ are expressed in terms of μ, σ as follows:

$$\begin{aligned} \zeta &= \sqrt{\ln\left(\frac{\sigma^2}{\mu^2} + 1\right)} \\ \lambda &= \ln \mu - \frac{\zeta^2}{2} \end{aligned} \tag{2.44}$$

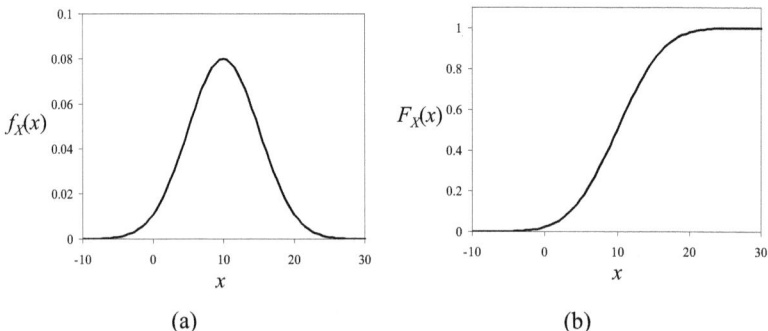

(a)   (b)

Figure 2.1: The PDF (a) and CDF (b) of a normal random variable with $\mu = 10$ and $\sigma = 5$.

2.2 Random variables

The lognormal distribution is an asymmetric distribution with positive skewness (i.e. $\gamma > 0$). It is usually applied for the description of variables that can only take positive values or in the case when experimental data sets exhibit a significant skewness. In Figure 2.2 the PDF of a lognormal random variable is compared to the one of a normal random variable with the same mean and standard deviation.

With the addition of a third parameter x_0, defining a lower bound in the support of x, the shifted lognormal random variable $X \sim LN(\lambda, \zeta, x_0)$ can be defined with the following PDF and CDF, respectively:

$$f_X(x) = \frac{1}{\zeta(x-x_0)} \varphi\left(\frac{\ln(x-x_0) - \lambda}{\zeta}\right) \quad x \in (x_0, +\infty) \tag{2.45}$$

$$F_X(x) = \Phi\left(\frac{\ln(x-x_0) - \lambda}{\zeta}\right) \tag{2.46}$$

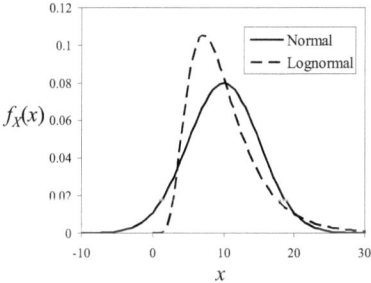

Figure 2.2: The lognormal PDF compared to the normal PDF for $\mu = 10$ and $\sigma = 5$.

Uniform distribution

A uniform random variable $X \sim U(a, b)$ is defined by the following PDF:

$$f_X(x) = \begin{cases} \dfrac{1}{b-a} & \text{for } x \in [a, b] \\ 0 & \text{otherwise} \end{cases} \tag{2.47}$$

whereby $b > a$. Its CDF is:

$$F_X(x) = \begin{cases} \dfrac{x-a}{b-a} & \text{for } x \in [a,b] \\ 0 & \text{otherwise} \end{cases} \qquad (2.48)$$

Its mean and standard deviation are expressed in terms of a, b as follows:

$$\begin{aligned} \mu &= \frac{a+b}{2} \\ \sigma &= \frac{b-a}{2\sqrt{3}} \end{aligned} \qquad (2.49)$$

Its skewness coefficient is $\gamma = 0$ and its kurtosis coefficient is $\kappa = 1.8$.

The uniform distribution assumes the same probability density for any value in the interval $[a, b]$. It can be used for the conservative probabilistic description of quantities for which bounds are given, but no specific trend can be assumed.

Beta distribution

A Beta-distributed random variable $X \sim Bet(q, r, a, b)$ is a four-parameter random variable with the following PDF and CDF, respectively:

$$f_X(x) = \begin{cases} \dfrac{(x-a)^{q-1}(b-x)^{r-1}}{B(q,r)(b-a)^{q+r-1}} & \text{for } x \in [a,b] \\ 0 & \text{otherwise} \end{cases} \qquad (2.50)$$

$$F_X(x) = \begin{cases} \dfrac{B\left(\dfrac{x-a}{b-a}; q, r\right)}{B(q,r)} & \text{for } x \in [a,b] \\ 0 & \text{otherwise} \end{cases} \qquad (2.51)$$

2.2 Random variables

whereby $b > a$ and $B(q, r)$, $B(x; q, r)$ are the Beta function and incomplete Beta function respectively, defined as follows:

$$B(q,r) = \int_0^1 z^{q-1}(1-z)^{r-1}\,dz \qquad (2.52)$$

$$B(x;q,r) = \int_0^x z^{q-1}(1-z)^{r-1}\,dz \qquad (2.53)$$

The mean and standard deviation are expressed as follows:

$$\begin{aligned} \mu &= \frac{ar+bq}{q+r} \\ \sigma &= \frac{b-a}{q+r}\sqrt{\frac{qr}{q+r+1}} \end{aligned} \qquad (2.54)$$

The Beta distribution is a very flexible distribution that can be used for the probabilistic description of bounded quantities. Figure 2.3 shows the Beta PDF for two different random variables with the same mean and standard deviation, but different bounds.

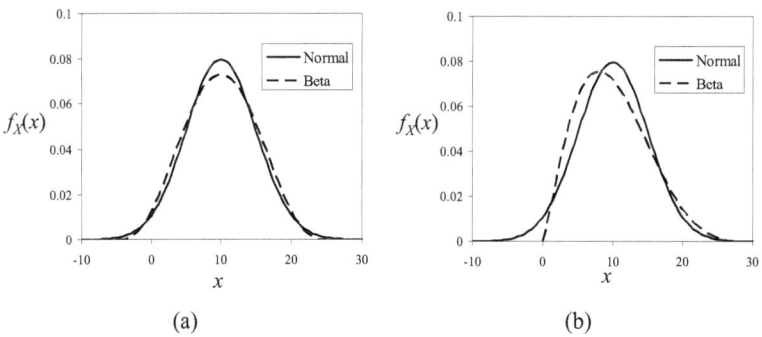

Figure 2.3: The Beta PDF compared to the normal PDF for $\mu = 10$ and $\sigma = 5$ and bounds (a) $a = -5$, $b = 25$ and (b) $a = 0$, $b = 30$.

Gumbel distribution

A random variable X follows a Gumbel or type I largest value distribution, i.e. $X \sim Gmb(u, \alpha)$, if its CDF has the following form:

$$F_X(x) = \exp\{-\exp[-\alpha(x-u)]\} \tag{2.55}$$

Its PDF is as follows:

$$f_X(x) = \alpha \exp\{-\alpha(x-u) - \exp[-\alpha(x-u)]\} \tag{2.56}$$

Its mean and standard deviation are expressed in terms of u, α as follows:

$$\begin{aligned} \mu &= u + \frac{\gamma}{\alpha} \\ \sigma &= \frac{\pi}{\sqrt{6}\alpha} \end{aligned} \tag{2.57}$$

where $\gamma \approx 0.577$ is Euler's constant.

The Gumbel distribution, derived from extreme value theory, models the distribution of the maximum value among a number of samples of any distribution with an exponential tail (e.g. the Gaussian distribution). It is commonly used for the modeling of environmental loads, such as winds and earthquakes.

Weibull distribution

A random variable X follows a Weibull distribution, i.e. $X \sim Wbl(u, k)$, if its CDF has the following form:

$$F_X(x) = \begin{cases} 1 - \exp\left[-\left(\frac{x}{u}\right)^k\right] & \text{for } x \in (0, +\infty) \\ 0 & \text{otherwise} \end{cases} \tag{2.58}$$

Its PDF is as follows:

$$f_X(x) = \begin{cases} \dfrac{k}{u}\left(\dfrac{x}{u}\right)^{k-1} \exp\left[-\left(\dfrac{x}{u}\right)^k\right] & \text{for } x \in (0, +\infty) \\ 0 & \text{otherwise} \end{cases} \qquad (2.59)$$

Its mean and standard deviation are expressed in terms of u, α as follows:

$$\mu = u\Gamma\left(1+\dfrac{1}{k}\right)$$
$$\sigma = u\sqrt{\Gamma\left(1+\dfrac{2}{k}\right) - \Gamma^2\left(1+\dfrac{1}{k}\right)} \qquad (2.60)$$

where $\Gamma(.)$ is the Gamma function, defined as:

$$\Gamma(k) = \int_0^\infty z^{k-1} \exp(-z)\, dz \qquad (2.61)$$

The Weibull distribution is identical to the type III smallest value distribution with a zero lower bound. Like the Gumbel distribution, it is derived from extreme value theory. It can be used to model the distribution of the minimum value of samples from most of the commonly used distributions with a zero lower bound. The PDFs of a Gumbel and a Weibull random variable with the same mean and standard deviation are shown in Figure 2.4.

The special case of a Weibull distribution with $k = 2$ results in a one-parameter distribution, called Rayleigh distribution $Ray(u)$. The Rayleigh distribution is commonly used for the probabilistic description of peaks in a narrow-banded stationary Gaussian process. For example, maximum wave heights can be modeled by a Rayleigh distribution.

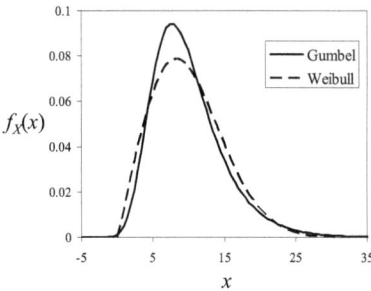

Figure 2.4: PDFs of commonly used extreme value distributions for $\mu = 10$ and $\sigma = 5$.

2.3 Random vectors

2.3.1 Joint distribution and joint moments

A random vector $\mathbf{X}$ is a mapping of the form $\mathbf{X} : S \rightarrow \mathbb{R}^n$, where n is the size of the vector. The components of the random vector are random variables, i.e. $\mathbf{X} = [X_1, \ldots, X_n]^T$, where $[.]^T$ denotes the transpose operator. The vector $\mathbf{X}$ can be completely defined by its joint CDF:

$$F_{\mathbf{X}}(\mathbf{x}) = F_{X_1,\ldots,X_n}(x_1,\ldots,x_n) = P(X_1 \leq x_1 \cap \ldots \cap X_n \leq x_n) \qquad (2.62)$$

The corresponding joint PDF can be obtained by differentiation of the joint CDF as follows:

$$f_{\mathbf{X}}(\mathbf{x}) = f_{X_1,\ldots,X_n}(x_1,\ldots,x_n) = \frac{\partial^n F_{X_1,\ldots,X_n}(x_1,\ldots,x_n)}{\partial x_1 \ldots \partial x_n} \qquad (2.63)$$

The joint PDF and CDF have the following normalization properties:

$$\lim_{x_i \to -\infty} F_{X_1,\ldots,X_n}(x_1,\ldots,x_i,\ldots,x_n) = 0 \qquad \forall i : 1 \leq i \leq n \qquad (2.64)$$

$$\lim_{(x_1,\ldots,x_n) \to (\infty,\ldots,\infty)} F_{X_1,\ldots,X_n}(x_1,\ldots,x_n) = 1 \qquad (2.65)$$

2.3 Random vectors

$$\int_{D_\mathbf{X}} f_\mathbf{X}(\mathbf{x})\, d\mathbf{x} = \int_{-\infty}^{\infty}\ldots\int_{-\infty}^{\infty} f_{X_1,\ldots,X_n}(x_1,\ldots,x_n)\, dx_1\ldots dx_n = 1 \qquad (2.66)$$

whereby $D_\mathbf{X} = \mathbb{R}^n$ and $d\mathbf{x} = dx_1\ldots dx_n$. The marginal distribution of any component random variable X_i ($1 \leq i \leq n$) is obtained by integrating the joint PDF over all remaining components:

$$f_{X_i}(x_i) = \int_{D_\mathbf{X}^{-1}} f_\mathbf{X}(\mathbf{x})\, d\mathbf{x}_{-i} \qquad (2.67)$$

where $D_\mathbf{X}^{-1} = \mathbb{R}^{n-1}$ and $d\mathbf{x}_{-i} = dx_1\ldots dx_{i-1}\, dx_{i+1}\ldots dx_n$. The joint PDF of two components X_i, X_j can be obtained in a similar way:

$$f_{X_i,X_j}(x_i,x_j) = \int_{D_\mathbf{X}^{-2}} f_\mathbf{X}(\mathbf{x})\, d\mathbf{x}_{-ij} \qquad (2.68)$$

The conditional PDF of any vector containing a subset of the components of the random vector $\mathbf{X}$, e.g. $\mathbf{X}_k = [X_1,\ldots, X_k]^\mathrm{T}$, given the joint PDF of the remaining components, $\mathbf{X}_{n-k} = [X_{k+1},\ldots, X_n]^\mathrm{T}$, is defined in analogy with Eq. (2.12), as follows:

$$f_{\mathbf{X}_k|\mathbf{X}_{n-k}}(\mathbf{x}_k\,|\,\mathbf{x}_{n-k}) = \frac{f_{\mathbf{X}_k,\mathbf{X}_{n-k}}(\mathbf{x}_k,\mathbf{x}_{n-k})}{f_{\mathbf{X}_{n-k}}(\mathbf{x}_{n-k})} = \frac{f_\mathbf{X}(\mathbf{x})}{f_{\mathbf{X}_{n-k}}(\mathbf{x}_{n-k})} \qquad (2.69)$$

We can then express the joint PDF $f_\mathbf{X}(\mathbf{x})$ by applying a generalized multiplication rule analogous to the one given in Eq. (2.14):

$$\begin{aligned} f_\mathbf{X}(\mathbf{x}) &= f_{X_1,\ldots,X_n}(x_1,\ldots,x_n) \\ &= f_{X_1|X_2,\ldots,X_n}(x_1\,|\,x_2,\ldots,x_n)\ldots f_{X_{n-1}|X_n}(x_{n-1}\,|\,x_n) f_{X_n}(x_n) \\ &= f_{X_n|X_1,\ldots,X_{n-1}}(x_n\,|\,x_1,\ldots,x_{n-1})\ldots f_{X_2|X_1}(x_2\,|\,x_1) f_{X_1}(x_1) \end{aligned} \qquad (2.70)$$

Eq. (2.70) implies that the joint PDF can be constructed if all the conditional PDFs are known. Note that the analyst is free to choose which conditional PDFs to estimate in order to obtain the joint PDF of a random vector, as long as the definition of conditional PDF is correctly applied in the construction of the multiplication rule.

Two random variables X_1, X_2 are said to be statistically independent if:

$$f_{X_1|X_2}(x_1|x_2) = f_{X_1}(x_1) \tag{2.71}$$

Eq. (2.71) implies that if X_1, X_2 are statistically independent then:

$$f_{X_1,X_2}(x_1,x_2) = f_{X_1}(x_1) f_{X_2}(x_2) \tag{2.72}$$

We can then obtain the joint PDF of a vector **X** of jointly statistically independent random variables as follows:

$$f_{\mathbf{X}}(\mathbf{x}) = f_{X_1,\ldots,X_n}(x_1,\ldots,x_n) = f_{X_1}(x_1)\ldots f_{X_n}(x_n) = \prod_{i=1}^{n} f_{X_i}(x_i) \tag{2.73}$$

Consider a function g(**X**) of the random vector **X**. The mathematical expectation E[g(**X**)] is defined in analogy with the definition given in Eq. (2.28) as follows:

$$E[g(\mathbf{X})] = \int_{D_{\mathbf{X}}} g(\mathbf{x}) f_{\mathbf{X}}(\mathbf{x}) d\mathbf{x}$$
$$= \int_{-\infty}^{\infty} \ldots \int_{-\infty}^{\infty} g(x_1,\ldots,x_n) f_{X_1,\ldots,X_n}(x_1,\ldots,x_n) dx_1 \ldots dx_n \tag{2.74}$$

The covariance Cov[X_i, X_j] of the random variables X_i, X_j is defined as follows:

$$\text{Cov}[X_i, X_j] = E\left[(X_i - \mu_{X_i})(X_j - \mu_{X_j})\right] \tag{2.75}$$

The dimensionless correlation coefficient of X_i, X_j is defined by normalizing the covariance by the standard deviations of the two random variables:

$$\rho_{X_i,X_j} = \frac{\text{Cov}[X_i, X_j]}{\sigma_{X_i} \sigma_{X_j}} \qquad \rho_{X_i,X_j} \in [-1,1] \tag{2.76}$$

The covariance and correlation coefficient are measures of the linear dependence of two random variables. Two random variables X_i, X_j are said to be uncorrelated if:

$$\text{Cov}[X_i, X_j] = 0 \tag{2.77}$$

It can be easily shown that if two random variables are statistically independent then they are also uncorrelated. Note that the reverse does not necessarily hold. Two random variables X_i, X_j are said to be orthogonal if:

$$\text{E}[X_i X_j] = 0 \tag{2.78}$$

The mean value vector $\boldsymbol{\mu}_\mathbf{X}$ of a random vector $\mathbf{X}$ is defined as the vector containing the mean value of each component random variable:

$$\boldsymbol{\mu}_\mathbf{X} = \left[\mu_{X_1}, \ldots, \mu_{X_n}\right]^T \tag{2.79}$$

The covariance matrix $\boldsymbol{\Sigma}_{\mathbf{XX}}$ and the correlation coefficient matrix $\mathbf{R}_{\mathbf{XX}}$ are square symmetric and positive definite matrices, defined as:

$$\boldsymbol{\Sigma}_{\mathbf{XX}} = \left[\text{Cov}[X_i, X_j]\right]_{n \times n} \tag{2.80}$$

$$\mathbf{R}_{\mathbf{XX}} = \left[\rho_{X_i, X_j}\right]_{n \times n} \tag{2.81}$$

We also define the diagonal matrix $\mathbf{D}_\mathbf{X}$ containing the standard deviation of each component random variable:

$$\mathbf{D}_\mathbf{X} = \text{diag}\left[\sigma_{X_i}\right]_{n \times n} \tag{2.82}$$

The covariance and correlation coefficient matrices satisfy the following relation:

$$\boldsymbol{\Sigma}_{\mathbf{XX}} = \mathbf{D}_\mathbf{X} \mathbf{R}_{\mathbf{XX}} \mathbf{D}_\mathbf{X} \tag{2.83}$$

2.3.2 Transformation of random vectors

Consider a random vector $\mathbf{X}$ with known joint PDF $f_\mathbf{X}(\mathbf{x})$ and a random vector $\mathbf{Y}$ whose components Y_i are related to the components of $\mathbf{X}$ by known functions, i.e.

$$y_i = g_i(x_1,\ldots,x_n) \qquad \forall i : 1 \le i \le n \qquad (2.84)$$

If we assume that the above defined mapping is one-to-one, then the joint PDF of **Y** is derived by requiring preservation of the probability content, as follows:

$$f_\mathbf{Y}(\mathbf{y}) = f_\mathbf{X}(\mathbf{x}) |\det \mathbf{J}_{x,y}| \qquad (2.85)$$

where $\mathbf{J}_{x,y}$ is the Jacobian matrix of the transformation, defined as follows:

$$\mathbf{J}_{x,y} = \left[\frac{\partial x_i}{\partial y_j} \right]_{n \times n} \qquad (2.86)$$

2.3.3 The multinormal distribution

Gaussian or normal random vectors have the unique property that they are completely defined by their mean value vector $\boldsymbol{\mu}_\mathbf{X}$ and covariance matrix $\boldsymbol{\Sigma}_{\mathbf{XX}}$. In addition, it can be shown that if a distribution is jointly Gaussian, then all lower order distributions as well as all conditional distributions are also Gaussian. Moreover, any linear mapping of a Gaussian random vector will result in another Gaussian vector, which will be jointly Gaussian with the original vector. The joint PDF of a Gaussian random vector **X** is as follows:

$$f_\mathbf{X}(\mathbf{x}) = \varphi_n(\mathbf{x} - \boldsymbol{\mu}_\mathbf{X}, \boldsymbol{\Sigma}_{\mathbf{XX}})$$
$$= \frac{1}{(2\pi)^{n/2} (\det \boldsymbol{\Sigma}_{\mathbf{XX}})^{1/2}} \exp\left[-\frac{1}{2}(\mathbf{x} - \boldsymbol{\mu}_\mathbf{X})^T \boldsymbol{\Sigma}_{\mathbf{XX}}^{-1}(\mathbf{x} - \boldsymbol{\mu}_\mathbf{X}) \right] \qquad (2.87)$$

where $\varphi_n(\mathbf{z}, \mathbf{R})$ denotes the n-dimensional standard normal PDF, associated with the vector **Z** of standard normal random variables with correlation coefficient matrix **R**:

$$\varphi_n(\mathbf{z}, \mathbf{R}) = \frac{1}{(2\pi)^{n/2} (\det \mathbf{R})^{1/2}} \exp\left[-\frac{1}{2} \mathbf{z}^T \mathbf{R}^{-1} \mathbf{z} \right] \qquad (2.88)$$

It should be noted that if a random vector is jointly normally distributed and its components are pair-wise uncorrelated, then they are also pair-wise independent. In the case where a random vector **U** consists of independent standard normal random variables, the joint PDF, denoted by $\varphi_n(\mathbf{u})$, reads:

$$\varphi_n(\mathbf{u}) = \prod_{i=1}^{n} \varphi(u_i) = \frac{1}{(2\pi)^{n/2}} \exp\left[-\frac{1}{2}\mathbf{u}^T \mathbf{u}\right] \tag{2.89}$$

The Gaussian distribution is widely used for modeling real problems because it often provides a good approximation of measured data. This can be explained by the central limit theorem, which states that if a random variable X is defined as the sum of a large number of random variables $\{Z_1,\ldots, Z_n\}$, then X is approximately Gaussian under some restrictions on the joint distribution of $\{Z_1,\ldots, Z_n\}$. A sufficient restriction is that $\{Z_1,\ldots, Z_n\}$ are independent and identically distributed. Assuming that $\{Z_1,\ldots, Z_n\}$ have expected values μ and variances σ^2, then the random variable X, defined as:

$$X = \frac{1}{n}\sum_{i=1}^{n} Z_i \tag{2.90}$$

converges in distribution to the Gaussian distribution with mean value μ and variance σ^2/n. A similar statement of the theorem exists for the sum of a large number of random vectors. Also, it is well known that the central limit theorem does not require complete independence between the random variables $\{Z_1,\ldots, Z_n\}$. The key restriction is that the variables are poorly correlated and that none of the variables has a dominant contribution to the total variance. In practical situations, the validity of the central limit theorem cannot be proven, since usually there is not enough information on the joint distribution of the random variables, contributing to the quantity of interest. It should be noted that the Gaussian distribution has significant limitations, such as its symmetry and the fact that the outcome space includes negative values. The following section deals with the case where the marginal distributions of a random vector are modeled by non-Gaussian distributions.

2.3.4 The Nataf model

In many cases, the probabilistic information of a random vector **X** is given in terms of the marginal distributions $F_{X_i}(x_i)$ and the covariance matrix Σ_{XX}. An approximation of the joint PDF can then be obtained using the Nataf distribution [85]. This can be achieved by assuming that the random variables Z_i derived from the following isoprobabilistic marginal transformation:

$$Z_i = \Phi^{-1}\left[F_{X_i}(X_i)\right] \qquad \forall i : 1 \leq i \leq n \tag{2.91}$$

form a jointly Gaussian random vector of standard normal random variables $\mathbf{Z} = [Z_i, ..., Z_n]^T$ with correlation coefficient matrix **R** [73]. Applying Eq. (2.85), the joint PDF of **X** is written as:

$$\begin{aligned}f_\mathbf{X}(\mathbf{x}) &= \varphi_n(\mathbf{z},\mathbf{R})|\det \mathbf{J}_{z,x}| \\ &= f_{X_1}(x_1) f_{X_2}(x_2) ... f_{X_n}(x_n) \frac{\varphi_n(\mathbf{z},\mathbf{R})}{\varphi(z_1)\varphi(z_2)...\varphi(z_n)}\end{aligned} \tag{2.92}$$

where $z_i = \Phi^{-1}\left[F_{X_i}(x_i)\right]$ and $f_{X_i}(x_i)$ are the given marginal PDFs. The elements ρ_{ij} of **R** are computed such that they match the given correlation coefficients ρ_{X_i,X_j} of **X**. This is achieved through the following integral relation:

$$\begin{aligned}\rho_{X_i,X_j} &= \frac{\text{Cov}\left[X_i,X_j\right]}{\sigma_{X_i}\sigma_{X_j}} \\ &= \int_{-\infty}^{\infty}\int_{-\infty}^{\infty}\left(\frac{x_i-\mu_{X_i}}{\sigma_{X_i}}\right)\left(\frac{x_j-\mu_{X_j}}{\sigma_{X_j}}\right) f_{X_i}(x_i) f_{X_j}(x_j) \frac{\varphi_2(z_i,z_j,\rho_{ij})}{\varphi(z_i)\varphi(z_j)} dx_i dx_j \\ &= \int_{-\infty}^{\infty}\int_{-\infty}^{\infty}\left(\frac{x_i-\mu_{X_i}}{\sigma_{X_i}}\right)\left(\frac{x_j-\mu_{X_j}}{\sigma_{X_j}}\right)\varphi_2(z_i,z_j,\rho_{ij}) dz_i dz_j\end{aligned} \tag{2.93}$$

The above relation is implicit in ρ_{ij} and can thus be solved iteratively. Alternatively, the empirical formulae given in [25], [73] relating ρ_{ij} to ρ_{X_i,X_j} for several combinations of distribution types can be applied. The

Nataf model is valid provided that the marginal CDFs $F_{X_i}(x_i)$ are continuous and strictly increasing and that the correlation coefficient matrix **R** derived from Eq. (2.93) is positive definite [73]. It should be noted that the latter condition does not necessarily hold, even though the correlation coefficient matrix $\mathbf{R_{xx}}$ is always positive definite. In such cases, the Nataf model is not applicable.

2.4 Random processes and random fields

In several engineering applications, the description of uncertain parameters using random variables can be insufficient. This is due to the fact that certain physical quantities are often expected to vary randomly in space or time. The probabilistic description of such quantities requires the consideration of random processes or fields (for a detailed view on the theory of random processes and fields, the reader is referred to [46] and [137]).

2.4.1 Basic definitions

A random (stochastic) process or field $X(t)$ is defined as a collection of random variables indexed by a continuous parameter $t \in T$, where T is some continuous set. This means that for any $t_i \in T$, $X(t_i)$ is a random variable. Then $x(t_i)$ will be an individual outcome of the corresponding sample space. Collecting the individual outcomes of all random variables $X(t_i)$, we get a realization of the random process, denoted here by $x(t)$. Figure 2.5 shows realizations of an example random process. Usually, a random process refers to the case where the set T denotes a continuous time interval. On the other hand, the notion of random field is used for describing spatially varying random quantities, i.e. $T = \Omega \subset \mathbb{R}^d$ ($d = 1, 2$ or 3). Hence, in the general case of some spatial domain Ω, the random field is denoted by $X(\mathbf{t})$, where $\mathbf{t}$ is now a location vector. In the following, we will use this more general notation, unless otherwise noted.

28   2 Modeling of uncertainties

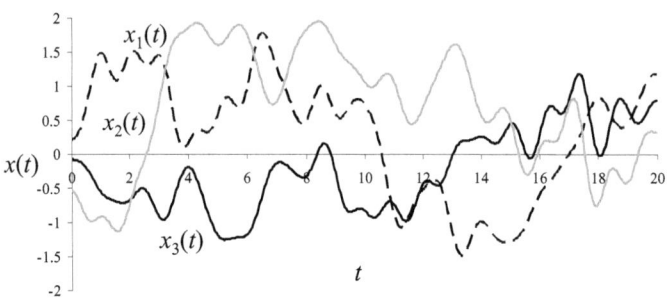

Figure 2.5:   Random realizations of a random process.

To completely define a random field, the joint PDF $f_{X(\mathbf{t}_1),...,X(\mathbf{t}_n)}(x_1,\mathbf{t}_1;...;x_n,\mathbf{t}_n)$ of the random variables $\{X(\mathbf{t}_1), X(\mathbf{t}_2),..., X(\mathbf{t}_n)\}$ for any $\{n, \mathbf{t}_1, \mathbf{t}_2, ..., \mathbf{t}_n\}$ must be specified. However, in most engineering applications, the joint PDF is approximated using second moment information and the marginal distribution, as will be discussed in Section 2.4.6.

Denoting by $f_{X(\mathbf{t})}(x,\mathbf{t})$ the marginal PDF of the random variable $X(\mathbf{t})$, for some $\mathbf{t} \in T$, the mean and variance functions of the field are defined as follows:

$$\mu_X(\mathbf{t}) = \mathrm{E}[X(\mathbf{t})] = \int_{-\infty}^{\infty} x f_{X(\mathbf{t})}(x,\mathbf{t}) dx \qquad (2.94)$$

$$\sigma_X^2(\mathbf{t}) = \mathrm{E}\left[(X(\mathbf{t})-\mu_X(\mathbf{t}))^2\right] = \int_{-\infty}^{\infty}(x-\mu_X(\mathbf{t}))^2 f_{X(\mathbf{t})}(x,\mathbf{t}) dx \qquad (2.95)$$

wherein $\sigma_X(\mathbf{t})$ is the standard deviation function. The second moment function or autocorrelation function is defined by:

$$R_{XX}(\mathbf{t}_1,\mathbf{t}_2) = \mathrm{E}[X(\mathbf{t}_1)X(\mathbf{t}_2)] = \int_{-\infty}^{\infty}\int_{-\infty}^{\infty} x_1 x_2 f_{X(\mathbf{t}_1)X(\mathbf{t}_2)}(x_1,\mathbf{t}_1;x_2,\mathbf{t}_2) dx_1 dx_2 \qquad (2.96)$$

where $f_{X(\mathbf{t}_1)X(\mathbf{t}_2)}(x_1,\mathbf{t}_1;x_2,\mathbf{t}_2)$ is the joint PDF of the random variables $\{X(\mathbf{t}_1), X(\mathbf{t}_2)\}$. The autocovariance function is defined as follows:

$$\Gamma_{XX}(\mathbf{t}_1,\mathbf{t}_2) = \text{Cov}\left[X(\mathbf{t}_1),X(\mathbf{t}_2)\right]$$
$$= \text{E}\left[\left(X(\mathbf{t}_1)-\mu_X(\mathbf{t}_1)\right)\left(X(\mathbf{t}_2)-\mu_X(\mathbf{t}_2)\right)\right] \quad (2.97)$$
$$= R_{XX}(\mathbf{t}_1,\mathbf{t}_2) - \mu_X(\mathbf{t}_1)\mu_X(\mathbf{t}_2)$$

A direct consequence of Eq. (2.97) is that the autocovariance and autocorrelation function are equal for the case of a zero mean random field. Setting $\mathbf{t}_1 = \mathbf{t}_2 = \mathbf{t}$ in Eq. (2.97), we obtain the variance function:

$$\sigma_X^2(\mathbf{t}) = \Gamma_{XX}(\mathbf{t},\mathbf{t}) = R_{XX}(\mathbf{t},\mathbf{t}) - \mu_X^2(\mathbf{t}) \quad (2.98)$$

where $R_{XX}(\mathbf{t},\mathbf{t}) = \text{E}\left[X(\mathbf{t})^2\right]$ is the mean square function. The autocovariance and autocorrelation functions are symmetric with respect to the diagonal plane $\mathbf{t}_1 = \mathbf{t}_2$. Also, they are positive semi-definite functions and have the following bounds:

$$\left|R_{XX}(\mathbf{t}_1,\mathbf{t}_2)\right| \leq \sqrt{R_{XX}(\mathbf{t}_1,\mathbf{t}_1)R_{XX}(\mathbf{t}_2,\mathbf{t}_2)} \quad (2.99)$$

$$\left|\Gamma_{XX}(\mathbf{t}_1,\mathbf{t}_2)\right| \leq \sigma_X(\mathbf{t}_1)\sigma_X(\mathbf{t}_2) \quad (2.100)$$

Finally, we define the autocorrelation coefficient function as follows:

$$\rho_{XX}(\mathbf{t}_1,\mathbf{t}_2) = \frac{\Gamma_{XX}(\mathbf{t}_1,\mathbf{t}_2)}{\sigma_X(\mathbf{t}_1)\sigma_X(\mathbf{t}_2)} \quad \rho_{XX}(\mathbf{t}_1,\mathbf{t}_2) \in [-1,1] \quad (2.101)$$

2.4.2 Homogeneous random fields

A random field $X(\mathbf{t})$ (resp. random process $X(t)$) is said to be strictly homogeneous (resp. strictly stationary) if its probabilistic structure is invariant to a shift in the parameter origin. Therefore, the following holds for any $\{n, \mathbf{t}_1, \mathbf{t}_2, \ldots, \mathbf{t}_n\}$ and $\mathbf{h} \in T$:

$$f_{X(\mathbf{t}_1),\ldots,X(\mathbf{t}_n)}(x_1,\mathbf{t}_1;\ldots;x_n,\mathbf{t}_n) = f_{X(\mathbf{t}_1+\mathbf{h}),\ldots,X(\mathbf{t}_n+\mathbf{h})}(x_1,\mathbf{t}_1+\mathbf{h};\ldots;x_n,\mathbf{t}_n+\mathbf{h}) \quad (2.102)$$

A direct consequence of Eq. (2.102) is that the marginal PDF of $X(t)$ is invariant in $\mathbf{t}$ and that the joint PDF of $\{X(\mathbf{t}_1), X(\mathbf{t}_2)\}$ is a function of the difference $\tau = \mathbf{t}_1 - \mathbf{t}_2$, i.e.:

$$f_{X(t)}(x,\mathbf{t}) \rightarrow f_{X(t)}(x) \tag{2.103}$$

$$f_{X(\mathbf{t}_1)X(\mathbf{t}_2)}(x_1,\mathbf{t}_1;x_2,\mathbf{t}_2) \rightarrow f_{X(\mathbf{t}_1)X(\mathbf{t}_2)}(x_1;x_2,\tau) \tag{2.104}$$

Hence, the mean and standard deviation functions are constant and the autocorrelation, autocovariance and autocorrelation coefficient functions are also functions of τ, i.e.

$$\mu_X(\mathbf{t}) \rightarrow \mu_X \tag{2.105}$$

$$\sigma_X(\mathbf{t}) \rightarrow \sigma_X \tag{2.106}$$

$$R_{XX}(\mathbf{t}_1,\mathbf{t}_2) \rightarrow R_{XX}(\tau) \tag{2.107}$$

$$\Gamma_{XX}(\mathbf{t}_1,\mathbf{t}_2) \rightarrow \Gamma_{XX}(\tau) \tag{2.108}$$

$$\rho_{XX}(\mathbf{t}_1,\mathbf{t}_2) \rightarrow \rho_{XX}(\tau) \tag{2.109}$$

If the probabilistic structure of a random field $X(\mathbf{t})$ (resp. random process $X(t)$) is invariant to a shift in the parameter origin only up to a second order [Eqs. (2.103)-(2.109)] then the field (resp. process) is said to be weakly or second-order homogeneous (resp. weakly stationary).

For a homogeneous random field, the bounds of $R_{XX}(\tau)$ and $\Gamma_{XX}(\tau)$ read:

$$|R_{XX}(\tau)| \leq R_{XX}(\mathbf{0}), \qquad R_{XX}(\mathbf{0}) = E\left[X(\mathbf{t})^2\right] = \text{const.} \tag{2.110}$$

$$|\Gamma_{XX}(\tau)| \leq \sigma_X^2 \tag{2.111}$$

The second-order moment functions of a real valued homogeneous field [Eqs. (2.107)-(2.109)] are even functions (i.e. symmetric with respect to the origin $\tau = 0$), e.g.:

$$R_{XX}(\tau) = R_{XX}(-\tau) \tag{2.112}$$

2.4 Random processes and random fields

and

$$R_{XX}(\tau_1,\ldots,-\tau_i,\ldots,\tau_d) = R_{XX}(-\tau_1,\ldots,\tau_i,\ldots,-\tau_d) \qquad \forall i: 1 \leq i \leq d \qquad (2.113)$$

where $\boldsymbol{\tau} = [\tau_1,\ldots, \tau_d]^T$. A homogeneous random field is said to be quadrant symmetric if its second-order functions are even with respect to each component of the vector $\boldsymbol{\tau}$ [137], e.g.

$$R_{XX}(\tau_1,\ldots,-\tau_i,\ldots,\tau_d) = R_{XX}(\tau_1,\ldots,\tau_i,\ldots,\tau_d) \qquad \forall i: 1 \leq i \leq d \qquad (2.114)$$

If the second-order functions of a random field depend on the distance between two points $\mathbf{t}_1$ and $\mathbf{t}_2$, i.e. the Euclidean norm $|\boldsymbol{\tau}| = |\mathbf{t}_1 - \mathbf{t}_2|$, then the corresponding correlation structure is said to be isotropic. We shall call a random field anisotropic, if each component of the vector $\boldsymbol{\tau}$ influences the second-order functions of the field in a different way.

An autocorrelation coefficient function is said to be fully separable if it can be expressed as a product of the autocorrelation coefficient functions of one-dimensional processes, i.e.:

$$\rho_{XX}(\boldsymbol{\tau}) = \rho_{XX}(\tau_1)\rho_{XX}(\tau_2)\ldots\rho_{XX}(\tau_d) \qquad (2.115)$$

Several models of autocorrelation coefficient functions for one-dimensional homogeneous fields have been proposed (e.g. see [137]). Common models are the exponential, squared exponential or Gaussian, and the triangular autocorrelation coefficient functions, defined respectively by:

$$\rho_A(\tau) = \exp\left(-\frac{|\tau|}{l_A}\right) \qquad (2.116)$$

$$\rho_B(\tau) = \exp\left(-\left(\frac{\tau}{l_B}\right)^2\right) \qquad (2.117)$$

$$\rho_C(\tau) = \begin{cases} 1 - \dfrac{|\tau|}{l_C} & \text{for } |\tau| \leq l_C \\ 0 & \text{otherwise} \end{cases} \qquad (2.118)$$

2 Modeling of uncertainties

The parameters l_A, l_B, l_C are the correlation lengths of the respective correlation models. A small correlation length signifies fast reduction of the correlation coefficient as the distance τ increases and thus a high variability in the random field realizations [see Figure 2.7(a) and (c)]. Conversely, large correlation lengths correspond to slowly varying realizations. Moreover, the limit case of an infinite correlation length can be modeled by one random variable. It is convenient to define a measure of the variability, independent of the adopted model. Such a measure is the scale of fluctuation θ [137], defined for a one-dimensional random field by:

$$\theta = \int_{-\infty}^{\infty} \rho_{XX}(\tau)d\tau = 2\int_{0}^{\infty} \rho_{XX}(\tau)d\tau \tag{2.119}$$

Figure 2.6 shows plots of the one-dimensional autocorrelation coefficient functions of Eqs (2.116)-(2.118) for a scale of fluctuation of $\theta = 10$. For the general case of a multi-dimensional homogeneous random field, the more general correlation parameter is defined by:

$$\eta = \int_{D_T} \rho_{XX}(\boldsymbol{\tau})d\boldsymbol{\tau} \tag{2.120}$$

where $D_T = \mathbb{R}_+^d$ and $d\boldsymbol{\tau} = d\tau_1\ldots d\tau_d$.

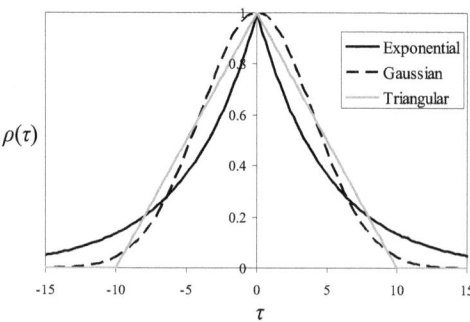

Figure 2.6: Autocorrelation coefficient functions for a scale of fluctuation of $\theta = 10$.

2.4 Random processes and random fields

For the special case where the autocorrelation coefficient function is fully separable, the correlation parameter is written as:

$$\eta = \prod_{i=1}^{d}\left[\int_{-\infty}^{\infty}\rho_{XX}(\tau_i)d\tau_i\right] = \theta_1\theta_2\ldots\theta_d \qquad (2.121)$$

where θ_i is the scale of fluctuation of the autocorrelation coefficient function $\rho_{XX}(\tau_i)$.

A second-order homogeneous random field is called ergodic if its second-order information can be obtained from a single realization of the field. For the one-dimensional case, we have:

$$\mu_X = \mathrm{E}[X(t)] = \langle x(t)\rangle = \lim_{T\to\infty}\frac{1}{2T}\int_{-T}^{T}x(t)dt \qquad (2.122)$$

$$\begin{aligned}R_{XX}(\tau) &= \mathrm{E}[X(t+\tau)X(t)] = \langle x(t+\tau)x(t)\rangle \\ &= \lim_{T\to\infty}\frac{1}{2T}\int_{-T}^{T}x(t+\tau)x(t)dt\end{aligned} \qquad (2.123)$$

where $\langle\cdot\rangle$ denotes the average operator over an interval T. The concept of ergodicity is of practical importance, since it allows the estimation of the statistics of a random field from a single time or space record. Ergodicity is usually assumed in the absence of evidence of the contrary. Note that an ergodic field is always homogeneous, but the reverse does not necessarily hold.

For a one-dimensional homogeneous field, the power spectral density function $\Phi_{XX}(\omega)$ gives the distribution of the mean energy of the field in the frequency domain. The autocorrelation function and the power spectral density function form a Fourier transform pair, i.e.

$$\Phi_{XX}(\omega) = \frac{1}{2\pi}\int_{-\infty}^{\infty}R_{XX}(\tau)e^{-i\omega\tau}d\tau = \frac{1}{\pi}\int_{0}^{\infty}R_{XX}(\tau)\cos\omega\tau\,d\tau \qquad (2.124)$$

$$R_{XX}(\tau) = \int_{-\infty}^{\infty}\Phi_{XX}(\omega)e^{i\omega\tau}d\omega = 2\int_{0}^{\infty}\Phi_{XX}(\omega)\cos\omega\tau\,d\omega \qquad (2.125)$$

where the second equality in both equations holds due to the symmetry of $R_{XX}(\tau)$, which implies that the power spectral density function $\Phi_{XX}(\omega)$ is symmetric about $\omega = 0$. Figure 2.7(a) and (b) show the exponential autocorrelation function and corresponding power spectral density function for different values of the scale of fluctuation. Eqs (2.124) and (2.125) are the Wiener-Khinchine relations and can be generalized for a random field of arbitrary dimension d, as follows:

$$\Phi_{XX}(\boldsymbol{\omega}) = \frac{1}{\pi^d} \int_{D_T} R_{XX}(\boldsymbol{\tau}) \cos \boldsymbol{\omega}^T \boldsymbol{\tau} \, d\boldsymbol{\tau} \tag{2.126}$$

$$R_{XX}(\boldsymbol{\tau}) = 2^d \int_{D_T} \Phi_{XX}(\boldsymbol{\omega}) \cos \boldsymbol{\omega}^T \boldsymbol{\tau} \, d\boldsymbol{\omega} \tag{2.127}$$

where $D_T = \mathbb{R}_+^d$, $d\boldsymbol{\tau} = d\tau_1 \ldots d\tau_d$, $\boldsymbol{\omega} = [\omega_1, \ldots, \omega_d]^T$ and $d\boldsymbol{\omega} = d\omega_1 \ldots d\omega_d$. For the special case of a quadrant symmetric random field, the Wiener-Khinchine relations read:

$$\Phi_{XX}(\boldsymbol{\omega}) = \frac{1}{\pi^d} \int_{D_T} R_{XX}(\boldsymbol{\tau}) \cos \omega_1 \tau_1 \ldots \cos \omega_d \tau_d \, d\boldsymbol{\tau} \tag{2.128}$$

$$R_{XX}(\boldsymbol{\tau}) = 2^d \int_{D_T} \Phi_{XX}(\boldsymbol{\omega}) \cos \omega_1 \tau_1 \ldots \cos \omega_d \tau_d \, d\boldsymbol{\omega} \tag{2.129}$$

2.4.3 Derivatives of random fields

The mean square derivative of a one-dimensional random process or field $X(t)$, denoted by $\dot{X}(t)$, is defined as follows:

$$\dot{X}(t) = \frac{d}{dt} X(t) = \operatorname*{l.i.m.}_{\Delta t \to 0} \frac{X(t + \Delta t) - X(t)}{\Delta t} \tag{2.130}$$

where l.i.m. denotes the limit in mean square. The limit in Eq. (2.130) exists, i.e. $\dot{X}(t)$ exists in the mean square sense, if the autocorrelation function $R_{XX}(t_1, t_2)$ is twice differentiable at $t_1 = t_2 = t$.

2.4 Random processes and random fields 35

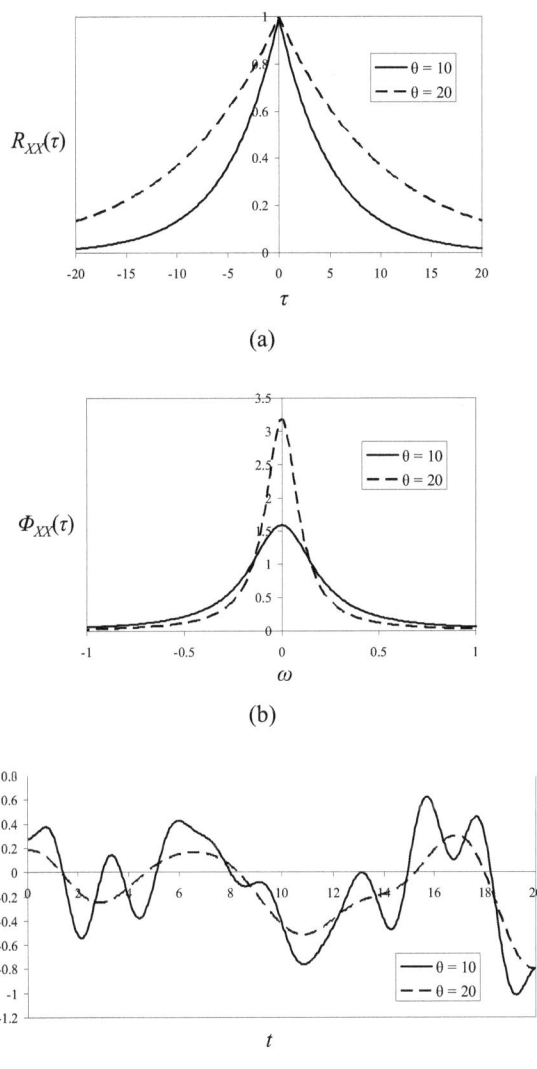

Figure 2.7: Autocorrelation function (a), power spectral density function (b) and corresponding realizations (c) of a one-dimensional standard normal ($\mu_X = 0$, $\sigma_X = 1$) stationary random process with exponential correlation model. Influence of the scale of fluctuation θ.

In this case, the mean function of $\dot{X}(t)$ is given by:

$$\mu_{\dot{X}}(t) = \mathrm{E}\left[\dot{X}(t)\right] = \frac{d}{dt}\mu_X(t) \tag{2.131}$$

The cross correlation between $X(t)$ and $\dot{X}(t)$ is computed as follows:

$$\begin{aligned} R_{X\dot{X}}(t_1,t_2) &= \mathrm{E}\left[X(t_1)\dot{X}(t_2)\right] = \lim_{\Delta t_2 \to 0} \mathrm{E}\left[X(t_1)\frac{X(t_2+\Delta t_2)-X(t_2)}{\Delta t_2}\right] \\ &= \lim_{\Delta t_2 \to 0} \mathrm{E}\left[\frac{R_{XX}(t_1,t_2+\Delta t_2)-R_{XX}(t_1,t_2)}{\Delta t_2}\right] = \frac{\partial R_{XX}(t_1,t_2)}{\partial t_2} \end{aligned} \tag{2.132}$$

Similarly, we have:

$$R_{\dot{X}X}(t_1,t_2) = \frac{\partial R_{XX}(t_1,t_2)}{\partial t_1} \tag{2.133}$$

$$R_{\dot{X}\dot{X}}(t_1,t_2) = \frac{\partial^2 R_{XX}(t_1,t_2)}{\partial t_1 \partial t_2} \tag{2.134}$$

Furthermore, if the process $X(t)$ is stationary with $\tau = t_1 - t_2$ then the mean, cross correlation and autocorrelation functions of its derivative process read:

$$\mu_{\dot{X}}(t) = 0 \tag{2.135}$$

$$R_{X\dot{X}}(\tau) = -\frac{dR_{XX}(\tau)}{d\tau} \tag{2.136}$$

$$R_{\dot{X}X}(\tau) = \frac{dR_{XX}(\tau)}{d\tau} \tag{2.137}$$

$$R_{\dot{X}\dot{X}}(\tau) = -\frac{d^2 R_{XX}(\tau)}{d\tau^2} \tag{2.138}$$

Eqs (2.135) and (2.138) show that the process $\dot{X}(t)$ is weakly stationary if $X(t)$ is weakly stationary. Since the autocorrelation function is an even function (see Section 2.4.2) and if its derivative process exists (note that as

2.4 Random processes and random fields

shown in Figure 2.6, the exponential and triangular models imply non differentiable random processes), it follows that at $\tau = 0$

$$\frac{dR_{XX}(0)}{d\tau} = 0 \tag{2.139}$$

Furthermore, we have

$$R_{X\dot{X}}(0) = E\left[X(t)\dot{X}(t)\right] = -\frac{dR_{XX}(0)}{d\tau} = 0 \tag{2.140}$$

$$R_{\dot{X}X}(0) = E\left[\dot{X}(t)X(t)\right] = \frac{dR_{XX}(0)}{d\tau} = 0 \tag{2.141}$$

Therefore, a stationary process and its derivative process are orthogonal and due to Eq. (2.135) uncorrelated for any $t \in T$.

It is useful to define the second-order information of the derivative of a stationary process in the frequency domain. Combining Eqs (2.136) and (2.125), we can express the cross correlation function of $X(t)$ and $\dot{X}(t)$ as follows:

$$R_{X\dot{X}}(\tau) = -\frac{dR_{XX}(\tau)}{d\tau} = -\int_{-\infty}^{\infty} i\omega \Phi_{XX}(\omega) e^{i\omega\tau} d\omega \tag{2.142}$$

Eq. (2.142) implies that the cross power spectral density function of the processes $X(t)$ and $\dot{X}(t)$ is given by:

$$\Phi_{X\dot{X}}(\omega) = -i\omega \Phi_{XX}(\omega) \tag{2.143}$$

Similarly, we have:

$$\Phi_{\dot{X}X}(\omega) = i\omega \Phi_{XX}(\omega) \tag{2.144}$$

$$\Phi_{\dot{X}\dot{X}}(\omega) = \omega^2 \Phi_{XX}(\omega) \tag{2.145}$$

The statistics of the derivative process given in Eqs (2.131)-(2.145) can be easily generalized for the partial derivative of the random field $X(\mathbf{t})$ with respect to t_i $\forall i : 1 \le i \le d$, defined as

$$\dot{X}_i(\mathbf{t}) = \frac{\partial}{\partial t_i} X(\mathbf{t}) \qquad (2.146)$$

For detailed derivations the reader is referred to [137].

2.4.4 Integrals of random fields

The mean square Riemann integral of a one-dimensional random process or field $X(t)$ over an interval $[a, b]$ is defined as follows:

$$I = \int_a^b X(t)\,dt = \mathrm{l.i.m.}_{n\to\infty} \Delta t \sum_{i=1}^{n} X(a + i\Delta t) \qquad (2.147)$$

where $\Delta t = (b - a)/n$. The quantity I is a random variable if the limits a, b are constant. It can be shown that the integral in Eq. (2.147) exists, i.e. $X(t)$ is mean square Riemann integrable, if the double integral of the autocorrelation function $R_{XX}(t_1, t_2)$ over the square domain $[a, b]^2$ is finite. If I exists then its first two moments are computed as follows:

$$\mu_I = \mathrm{E}[I] = \lim_{n\to\infty} \Delta t \sum_{i=1}^{n} \mu_X(a + i\Delta t) = \int_a^b \mu_X(t)\,dt \qquad (2.148)$$

$$\mathrm{E}[I^2] = \lim_{n\to\infty} \Delta t^2 \sum_{i=1}^{n}\sum_{j=1}^{n} R_{XX}(a+i\Delta t, a+j\Delta t) = \int_a^b\int_a^b R_{XX}(t_1,t_2)\,dt_1 dt_2 \qquad (2.149)$$

If the process $X(t)$ is stationary with $\tau = t_1 - t_2$, then the first two moments of I read:

$$\mu_I = (b - a)\mu_X \qquad (2.150)$$

$$\mathrm{E}[I^2] = \int_a^b\int_a^b R_{XX}(t_1 - t_2)\,dt_1 dt_2 \qquad (2.151)$$

The integral in Eq. (2.151) can be further simplified by changing the variables applying the mapping $T(t_1,t_2) = \langle s, s - \tau \rangle$. If we then integrate with respect to s and use the fact that R_{XX} is an even function the integral reads [137]:

2.4 Random processes and random fields

$$\mathrm{E}\left[I^2\right] = 2\int_0^{b-a} R_{XX}(\tau)\left[(b-a)-\tau\right]d\tau \qquad (2.152)$$

For detailed derivations the reader is referred to [137], wherein additional results are provided for integrals of homogeneous random fields in rectangular domains.

Consider now the case where we are interested in the integral of the random process, multiplied by a kernel function $g(t, s)$, i.e.

$$I(s) = \int_a^b X(t) g(t,s) dt \qquad (2.153)$$

For any value of s the integral in Eq. (2.153) is a random variable, i.e. $I(s)$ is a random process. Applying Eqs. (2.148) and (2.149), we have:

$$\mu_I(s) = \int_a^b \mu_X(t) g(t,s) dt \qquad (2.154)$$

$$R_{II}(s_1,s_2) = \int_a^b\int_a^b R_{XX}(t_1,t_2) g(t_1,s_1) g(t_2,s_2) dt_1 dt_2 \qquad (2.155)$$

The integral process $I(s)$ is important for establishing the input-output relationships in random vibrations analysis.

2.4.5 Gaussian random fields

A random field $X(\mathbf{t})$ is Gaussian if for any $\{n, \mathbf{t}_1, \mathbf{t}_2, \ldots, \mathbf{t}_n\}$ the random variables $\{X(\mathbf{t}_1), X(\mathbf{t}_2),\ldots, X(\mathbf{t}_n)\}$ are jointly Gaussian. Therefore, any linear mapping of a Gaussian field is also Gaussian, as well as jointly Gaussian with the original field. In particular, any derivative field $\dot{X}_i(\mathbf{t})$ of a Gaussian field $X(\mathbf{t})$ is Gaussian, due to the definition of the derivative process in Eq. (2.130). Moreover, $\dot{X}_i(\mathbf{t})$ is jointly Gaussian with $X(\mathbf{t})$. Similarly, any integral of a Gaussian process is a Gaussian random variable [or a Gaussian random field for integrals of the form of Eq. (2.153)] and is jointly Gaussian with the original field, due to the definition in Eq. (2.147).

In agreement with the definition of the Gaussian vector in Section 2.3.3, a Gaussian random field can be completely defined by its mean function $\mu_X(\mathbf{t})$ and either its autocovariance function $\Gamma_{XX}(\mathbf{t}_1,\mathbf{t}_2)$ or its autocorrelation function $R_{XX}(\mathbf{t}_1,\mathbf{t}_2)$. Alternatively, the correlation structure can be defined by the standard deviation function $\sigma_X(\mathbf{t})$ and the autocorrelation coefficient function $\rho_{XX}(\mathbf{t}_1,\mathbf{t}_2)$. The marginal PDF of the Gaussian random field is given by:

$$f_{X(\mathbf{t})}(x,\mathbf{t}) = \frac{1}{\sigma_X(\mathbf{t})}\varphi\left(\frac{x-\mu_X(\mathbf{t})}{\sigma_X(\mathbf{t})}\right)$$

$$= \frac{1}{\sigma_X(\mathbf{t})\sqrt{2\pi}}\exp\left[-\frac{(x-\mu_X(\mathbf{t}))^2}{2\sigma_X^2(\mathbf{t})}\right] \quad (2.156)$$

where $\varphi(.)$ as defined in Eq. (2.38) is the marginal PDF of the standardized Gaussian random field $U(\mathbf{t})$, defined as:

$$U(\mathbf{t}) = \frac{X(\mathbf{t})-\mu_X(\mathbf{t})}{\sigma_X(\mathbf{t})} \quad (2.157)$$

Clearly, the field $U(\mathbf{t})$ has zero mean and unit standard deviation. Moreover, the joint PDF of the random variables $\{X(\mathbf{t}_1), X(\mathbf{t}_2),\ldots, X(\mathbf{t}_n)\}$ for any $\{n, \mathbf{t}_1, \mathbf{t}_2, \ldots, \mathbf{t}_n\}$ is given by Eq. (2.87), whereby the mean value vector $\boldsymbol{\mu}_\mathbf{X}$ and covariance matrix $\boldsymbol{\Sigma}_\mathbf{XX}$ are calculated by:

$$\boldsymbol{\mu}_\mathbf{X} = \left[\mu_X(\mathbf{t}_1),\ldots,\mu_X(\mathbf{t}_n)\right]^T \quad (2.158)$$

$$\boldsymbol{\Sigma}_\mathbf{XX} = \left[\Gamma_{XX}(\mathbf{t}_i,\mathbf{t}_j)\right]_{n\times n} \quad (2.159)$$

If a Gaussian random field is homogeneous, it can be completely defined by its mean value μ_X, standard deviation σ_X and autocorrelation coefficient function $\rho_{XX}(\boldsymbol{\tau})$. Therefore, a weakly homogeneous Gaussian random field is also strictly homogeneous. Figure 2.7(c) shows realizations of a one-dimensional standard Gaussian homogeneous random field with corresponding exponential autocorrelation coefficient functions as shown in Figure 2.7(a).

2.4.6 Non-Gaussian random fields

In the general case, where the considered random field $X(\mathbf{t})$ is not Gaussian (or of some other special types), the corresponding joint PDF of the random variables $\{X(\mathbf{t}_1), X(\mathbf{t}_2),\ldots, X(\mathbf{t}_n)\}$ for any $\{n, \mathbf{t}_1, \mathbf{t}_2, \ldots, \mathbf{t}_n\}$ is in practice impossible to obtain. However, a class of non-Gaussian random fields with given marginal distribution $F_{X(\mathbf{t})}$ and second moment information can be defined by a nonlinear marginal transformation (translation) of an underlying Gaussian field $U(\mathbf{t})$ [44], of the form

$$X(\mathbf{t}) = g[U(\mathbf{t})] \qquad (2.160)$$

This class of non-Gaussian fields is called translation fields. A special case of translation fields is defined based on the Nataf distribution (see Section 2.3.4). For this case, the marginal transformation is given by:

$$X(\mathbf{t}) = F_{X(\mathbf{t})}^{-1}\{\Phi[U(\mathbf{t})], \mathbf{t}\} \qquad (2.161)$$

where $\Phi(.)$ is the standard normal CDF. Based on Eq. (2.161), the Gaussian field $U(\mathbf{t})$ has zero mean and unit variance. Moreover, its autocorrelation coefficient function $\rho_{UU}(\mathbf{t}_1,\mathbf{t}_2)$ is obtained by the following integral equation [72]:

$$\rho_{XX}(\mathbf{t}_1,\mathbf{t}_2) = \int_{-\infty}^{\infty}\int_{-\infty}^{\infty} \left(\frac{x(\mathbf{t}_1)-\mu_X(\mathbf{t}_1)}{\sigma_X(\mathbf{t}_1)}\right)\left(\frac{x(\mathbf{t}_2)-\mu_X(\mathbf{t}_2)}{\sigma_X(\mathbf{t}_2)}\right)\varphi_2[u_1,u_2,\rho_{UU}(\mathbf{t}_1,\mathbf{t}_2)]du_1 du_2 \qquad (2.162)$$

where $\mu_X(\mathbf{t})$, $\sigma_X(\mathbf{t})$ and $\rho_{XX}(\mathbf{t}_1,\mathbf{t}_2)$ are the mean, standard deviation and autocorrelation coefficient functions of $X(\mathbf{t})$. Eq. (2.162) can be solved iteratively for $\rho_{UU}(\mathbf{t}_1,\mathbf{t}_2)$. Alternatively, the empirical formulae given in [25], [73] relating $\rho_{UU}(\mathbf{t}_1,\mathbf{t}_2)$ to $\rho_{XX}(\mathbf{t}_1,\mathbf{t}_2)$ for common marginal distribution types can be applied.

A different type of translation field can be defined by expansion of the original field using one-dimensional Hermite polynomials with the underlying Gaussian field as argument ([104], [114]). In addition, it is shown in [101] that a non-Gaussian field can be defined by its marginal distribution and second-moment functions in conjunction with the spectral expansion methods described in Section 2.5.3, without the need for a translation from an equivalent Gaussian field. However, this approach requires additional considerations at the discretization level.

2.5 Discretization of random fields

A random field is an infinite set of random variables by definition, which makes its computational handling impossible. The discretization of the continuous field $X(\mathbf{t})$ consists in its approximation by a discrete $\hat{X}(\mathbf{t})$ defined by means of a finite set of random variables $\{X_1, X_2,..., X_n\}$. Several methods have been proposed for the discretization of random fields – a comprehensive review is given in [132]. These methods can be divided into the following three categories:

- Point discretization methods
- Average discretization methods
- Series expansion or spectral methods

For the case of a random field defined over an arbitrary shaped domain $\Omega \subset \mathbb{R}^d$ ($d \geq 2$), most methods require the splitting of this domain into a discrete assembly of individual elements Ω_d, which is usually referred to as the stochastic finite element (SFE) mesh (Figure 2.8). Note that in the case of point or average discretization methods, the SFE mesh is directly related to the random variables derived from the discretization, while for the series expansion methods, this relation is indirect. Moreover, it is shown in Section 2.5.5 that, in principle, series expansion methods may be applied without a discrete description of the actual domain of interest Ω, provided that Ω does not differ much from a corresponding rectangular embedded domain Ω_r (Figure 2.14).

2.5 Discretization of random fields

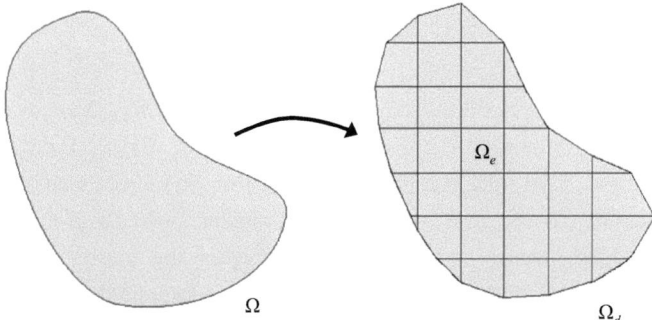

Figure 2.8: Representation of the domain Ω by a discrete domain Ω_d stochastic finite element mesh.

The discretization method should be able to approximate the random field with as few random variables as possible. To this end, it is useful to define an error measure for validation purposes. The error measure used in this study is the point-wise variance of the difference of the approximated field from the original field, divided by the variance of the original field, i.e.

$$err(\mathbf{t}) = \frac{\text{Var}\left[X(\mathbf{t}) - \hat{X}(\mathbf{t})\right]}{\text{Var}\left[X(\mathbf{t})\right]} \qquad (2.163)$$

The choice of an error measure based on a second-order function is well justified, in the case where the random field is defined using second-order information. In the next sections, a short overview of the different discretization techniques is provided.

2.5.1 Point discretization methods

In point discretization methods, the derived random variables $\{X_1, X_2, ..., X_n\}$ correspond to the values $\{X(\mathbf{t}_1), X(\mathbf{t}_2), ..., X(\mathbf{t}_n)\}$ of $X(\mathbf{t})$ at discrete points of the domain Ω. These points are based on the selection of a SFE mesh Ω_d, consisting of a finite number of elements Ω_e (Figure 2.8). The variables $\{X_1, X_2, ..., X_n\}$ are correlated random variables, with the following covariance matrix:

$$\Sigma_{XX} = \left[\Gamma_{XX}\left(t_i, t_j\right) \right]_{n \times n} \tag{2.164}$$

where $\Gamma_{XX}(t_1, t_2)$ is the autocovariance function of $X(t)$. Moreover, the marginal distribution of X_i coincides with the one of $X(t)$ for $t = t_i$. In the general case, where the marginal distribution of $X(t)$ is not Gaussian, the joint PDF of $\{X_1, X_2, \ldots, X_n\}$ can be approximated by the Nataf model (see Section 2.3.4), which is equivalent to applying the translation of Eq. (2.161) to the field $X(t)$. The approximated random field $\hat{X}(t)$ is then expressed as follows:

$$\hat{X}(t) = \sum_{i=1}^{n} X_i \varphi_i(t) \tag{2.165}$$

where $\{\varphi_1(t), \varphi_2(t), \ldots, \varphi_n(t)\}$ are deterministic functions. We can then compute the variance of the discretization error, as follows:

$$\operatorname{Var}\left[X(t) - \hat{X}(t)\right] = \sigma_X^2(t) + \sum_{i=1}^{n}\sum_{j=1}^{n} \varphi_i(t)\varphi_j(t)\operatorname{Cov}\left[X_i, X_j\right] \\ - 2\sum_{i=1}^{n}\varphi_i(t)\operatorname{Cov}\left[X(t), X_i\right] \tag{2.166}$$

In the following, the most commonly used point discretization methods are briefly discussed.

The midpoint method

In the midpoint method [24], the random variables $\{X_1, X_2, \ldots, X_n\}$ are chosen as the values of the field at the midpoint (center of gravity) t_c of each element Ω_e. Moreover, each function $\varphi_i(t)$ is chosen such that the value of the field is constant over each element of the SFE mesh, i.e.

$$\varphi_e(t) = \begin{cases} 1 & \text{if } t \in \Omega_e \\ 0 & \text{otherwise} \end{cases} \tag{2.167}$$

Therefore, each realization of the approximated field $\hat{X}(t)$ is piecewise constant, with discontinuities arising at the boundaries of each element. It is

2.5 Discretization of random fields

shown [24] that the midpoint method tends to over-represent the variability of the random field within each element.

The shape function method

The shape function method [75] resembles the deterministic linear finite element (FE) method, in the sense that it approximates the random field using the random variables $\{X_1, X_2,\ldots, X_n\}$ corresponding to the nodal points t_i of the SFE mesh. The deterministic functions $\varphi_i(t)$ are then chosen as the piecewise linear polynomials used as shape functions in FE analysis, defined such that:

$$\varphi_i(\mathbf{t}) = \begin{cases} 1 & \text{if } \mathbf{t} = \mathbf{t}_i \\ 0 & \text{if } \mathbf{t} = \mathbf{t}_j \ \forall j \neq i \end{cases} \quad (2.168)$$

A realization of the random field is obviously a piecewise linear continuous function, which is an advantage compared to the midpoint method.

The optimal linear estimation method

The optimal linear estimation method [72] may in principle be applied using the random variables corresponding to any set of points in the domain Ω. Without loss of generality, we will assume here that the nodal points t_i of the SFE mesh are used. The method starts with assuming that the random field is approximated as a linear function of its values at the points t_i, i.e.

$$\hat{X}(\mathbf{t}) = a(\mathbf{t}) + \sum_{i=1}^{n} X_i b_i(\mathbf{t}) \quad (2.169)$$

The functions $a(\mathbf{t})$ and $b_i(\mathbf{t})$ are then found by minimizing the variance error $\mathrm{Var}\left[X(\mathbf{t}) - \hat{X}(\mathbf{t})\right]$ at each point $\mathbf{t}$ subject to $\hat{X}(\mathbf{t})$ being an unbiased estimator of $X(\mathbf{t})$ in the mean, i.e.

$$\begin{array}{ll} \text{minimize} & \mathrm{Var}\left[X(\mathbf{t}) - \hat{X}(\mathbf{t})\right] \\ \text{subject to} & \mathrm{E}\left[X(\mathbf{t}) - \hat{X}(\mathbf{t})\right] = 0 \end{array} \quad (2.170)$$

2 Modeling of uncertainties

The solution of the optimization problem of Eq. (2.170) is [72]:

$$a(t) = \mu_X(t) - \mathbf{b}^T(t)\boldsymbol{\mu_X}$$
$$\mathbf{b}(t) = \boldsymbol{\Sigma}_{\mathbf{XX}}^{-1}\boldsymbol{\Sigma}_{X(t)\mathbf{X}} \qquad (2.171)$$

where $\mathbf{b}(t) = [b_1(t), b_2(t), ..., b_n(t)]^T$, $\mu_X(t)$ is the mean function of $X(t)$, $\boldsymbol{\mu_X}$ is the mean value vector of the random vector $\mathbf{X} = [X_1, X_2, ..., X_n]^T$, $\boldsymbol{\Sigma}_{\mathbf{XX}}$ is the covariance matrix of $\mathbf{X}$ and $\boldsymbol{\Sigma}_{X(t)\mathbf{X}}$ is the vector containing the covariances of $X(t)$ with the elements of $\mathbf{X}$, i.e.:

$$\boldsymbol{\Sigma}_{X(t)\mathbf{X}} = [\Gamma_{XX}(\mathbf{t}, \mathbf{t}_i)]_{n \times 1} \qquad (2.172)$$

The approximated field is then expressed as follows:

$$\hat{X}(\mathbf{t}) = \mu_X(\mathbf{t}) - \boldsymbol{\Sigma}_{X(t)\mathbf{X}}^T \boldsymbol{\Sigma}_{\mathbf{XX}}^{-1} \boldsymbol{\mu_X} + \sum_{i=1}^{n} X_i \left(\boldsymbol{\Sigma}_{\mathbf{XX}}^{-1} \boldsymbol{\Sigma}_{X(t)\mathbf{X}}\right)_i \qquad (2.173)$$

Eq. (2.173) implies that the field is approximated by a deterministic part and an expression identical to the one of Eq. (2.165), whereby:

$$\varphi_i(\mathbf{t}) = \left(\boldsymbol{\Sigma}_{\mathbf{XX}}^{-1} \boldsymbol{\Sigma}_{X(t)\mathbf{X}}\right)_i \qquad (2.174)$$

The variance of the error can be computed by substituting Eq. (2.174) to Eq. (2.166), which gives:

$$\mathrm{Var}\left[X(\mathbf{t}) - \hat{X}(\mathbf{t})\right] = \sigma_X^2(\mathbf{t}) - \boldsymbol{\Sigma}_{X(t)\mathbf{X}}^T \boldsymbol{\Sigma}_{\mathbf{XX}}^{-1} \boldsymbol{\Sigma}_{X(t)\mathbf{X}} \qquad (2.175)$$

The second term in Eq. (2.175) is identical to the variance of $\hat{X}(\mathbf{t})$. Therefore, the variance of the error is equal to the difference between the variances of $X(\mathbf{t})$ and $\hat{X}(\mathbf{t})$. Since the variance of the error is always positive, it follows that $\hat{X}(\mathbf{t})$ always under-estimates the variance of the original field.

It should be noted that the shape function and optimal linear estimation methods, as well as any other linear estimation method, are mainly applicable for the discretization of Gaussian fields, due to the linearity of the Gaussian distribution.

2.5.2 Average discretization methods

In average discretization methods, the random variables representing the random field are derived from a weighted averaging of the field over discrete parts of the domain Ω. These are chosen as the elements of the SFE mesh Ω_d, just as in point discretization methods. This category includes the spatial average method [137], [138] and the weighted integral method [28]-[30] (developed for applications of the perturbation-based stochastic finite element method). A brief description of the spatial average method is given next.

The Spatial average method

In the spatial average method [137], the original field is approximated in each SFE by the average of the field X_i over the element, i.e.

$$\hat{X}(\mathbf{t}) = X_i = \frac{1}{|\Omega_e|} \int_{\Omega_e} X(\mathbf{t}) \, d\Omega_e \qquad \mathbf{t} \in \Omega_e \qquad (2.176)$$

The mean value vector and covariance matrix of the derived random variables $\mathbf{X} = [X_1, X_2, \ldots, X_n]^T$ are evaluated as integrals of the mean and covariance function of the original field. Consider the case of a one-dimensional homogeneous random field $X(t)$. Its spatial average over an SFE of length T reads:

$$X_i = \frac{1}{T} \int_{t_i - T/2}^{t_i + T/2} X(t) \, dt \qquad (2.177)$$

where t_i is the midpoint of the SFE i. According to Eqs. (2.150) and (2.152), the mean and variance of X_i are as follows:

$$\mu_{X_i} = \mu_X \qquad (2.178)$$

$$\text{Var}[X_i] = \sigma_X^2 \frac{2}{T} \int_0^T \rho_{XX}(\tau)\left(1 - \frac{\tau}{T}\right) d\tau = \sigma_X^2 \, \gamma(T) \qquad (2.179)$$

where $\gamma(T) \in (0,1]$ is the variance function [137] and expresses the reduction of the variance of the original field under spatial averaging. The covariance of the random variables X_i, X_j, derived from spatial averaging over elements with length T and T', respectively, is given as follows [137]:

$$\text{Cov}\left[X_i, X_j\right] = \frac{1}{TT'} \frac{\sigma_X^2}{2} \sum_{k=0}^{3} (-1)^k T_k^2 \gamma(T_k) \qquad (2.180)$$

where T_k are distances that characterize the relative positions of the elements i, j [see Figure 2.9(a)]. The covariance of each random variable X_i with the actual field $X(t)$ for every point inside the SFE i can be computed by taking the limit $T' \to 0$ in Eq. (2.180) for the case where element i contains element j. We can therefore compute the point-wise variance of the discretization error inside each SFE as follows [137]:

$$\text{Var}\left[X(t) - \hat{X}(t)\right] = \sigma_X^2 + \sigma_X^2 \gamma(T) - 2\text{Cov}\left[X(t), X_i\right]$$

$$= \sigma_X^2 \left[1 + \gamma(T) - \frac{1}{T}\left(\frac{d(u^2\gamma(u))}{du}\bigg|_{u=t} + \frac{d(u^2\gamma(u))}{du}\bigg|_{u=T-t}\right)\right] \qquad (2.181)$$

It is shown [24] that the spatial average method tends to under-represent the variability of the random field within each element.

As shown in Eq. (2.179), the variance function $\gamma(T)$ depends on the respective correlation model. However, in some cases information about the correlation structure of $X(t)$ is limited to the scale of fluctuation θ [see Eq. (2.119)]. In such cases, the variance function can be approximated as follows [137]:

$$\gamma(T) = \begin{cases} 1 & \text{if } T \leq \frac{\theta}{2} \\ \frac{\theta}{T}\left(1 - \frac{\theta}{4T}\right) & \text{if } T > \frac{\theta}{2} \end{cases} \qquad (2.182)$$

2.5 Discretization of random fields   49

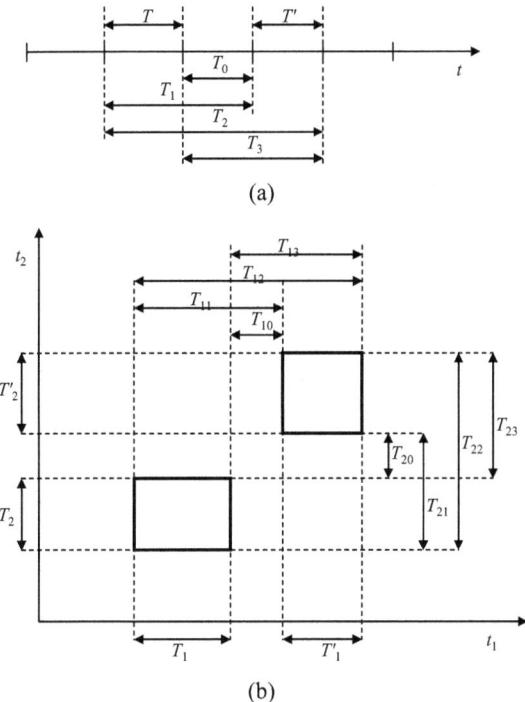

Figure 2.9: Distances that characterize the relative positions of the stochastic finite elements for the spatial average method in the one-dimensional (a), and two-dimensional (b) cases

Eq. (2.182) provides a good approximation of the correlation structure of wide-band models. This expression is frequently used to approximately account for the spatial variability of a random quantity, bypassing the discretization of the relevant random field (e.g. see [121]). The random field is thus modeled by one random variable with reduced variance, as obtained by Eq. (2.182) where T represents the size of the domain of interest.

Similar results can be obtained for multi-dimensional homogeneous fields $X(\mathbf{t})$, in the case where the spatial averaging is performed in rectangular domains [137]. In the two-dimensional case, the variance function is defined as follows:

50   2 Modeling of uncertainties

$$\gamma(T_1,T_2) = \frac{4}{T_1T_2^2}\int_0^{T_1}\int_0^{T_2} \rho_{XX}(\tau_1,\tau_2)\left(1-\frac{\tau_1}{T_2}\right)\left(1-\frac{\tau_2}{T_2}\right)d\tau_2 d\tau_1 \qquad (2.183)$$

whereby $T_1 \times T_2$ is the size of the rectangular element and quadrant symmetry is assumed. If the correlation structure is separable, the variance function is given as the product of the variance functions of one-dimensional fields. The covariance of the random variables corresponding to the spatial averages over two rectangular elements is obtained by a double sum over the characteristic distances in each dimension [see Figure 2.9(b)]. If the SFE mesh consists of elements of arbitrary shape, the elemental domains can be approximated by a collection of rectangular domains and the expression of the variance function in Eq. (2.183) can be used [137]. Also, non-homogeneous random fields can be modeled by assuming that the normalization of Eq. (2.157) results in a homogeneous field.

If the random field is Gaussian, then the random variables derived from spatial averaging will be jointly Gaussian (see Section 2.4.5). However, in the case of non-Gaussian random fields, the distribution of the local integral of the field is almost impossible to obtain. In such cases, a transformation of the form of Eq. (2.161) to an equivalent Gaussian field must be performed prior to the discretization.

2.5.3 Series expansion methods

Series expansion methods approximate the random field as a truncated series involving random variables and deterministic spatial functions, defined globally in the domain Ω. Each term in the series expansion as well as each random variable derived from the discretization have a global influence in the random realizations of the approximated field, unlike in point or average discretization methods, where the random variables correspond to local values of the field. An overview of the most important series expansion methods is given next.

2.5 Discretization of random fields

The spectral representation method

This method is based on the spectral representation theorem, which states that a weakly stationary process can be represented by a superposition of harmonics with random properties. Consider the zero-mean stationary random process $X(t)$ with power spectral density function $\Phi_{XX}(\omega)$ that is assumed to be negligible for frequencies $\omega > \omega_c$, $0 < \omega_c < \infty$, where ω_c is some cut-off frequency. The spectral representation of $X(t)$ reads:

$$\hat{X}(t) = \sum_{i=0}^{m} \sqrt{2\Phi_{XX}(\omega_i)\Delta\omega} \left(X_i \cos\omega_i t + Y_i \sin\omega_i t \right)$$
$$= \sum_{i=0}^{m} \sqrt{2\Phi_{XX}(\omega_i)\Delta\omega} A_i \cos(\omega_i t + \Psi_i) \qquad (2.184)$$

where $\omega_i = i\Delta\omega$, $\Delta\omega = \omega_c/m$. In the first expression of Eq. (2.184), the variables $\{X_i, Y_i\}$ are uncorrelated random variables with zero mean and unit variance. In the second expression, $A_i = (X_i + Y_i)^{1/2}$ and $\Psi_i = -\tan^{-1}(Y_i/X_i)$. If the process $X(t)$ is Gaussian, then the variables $\{X_i, Y_i\}$ are independent Gaussian variables, $\{A_i\}$ are independent Rayleigh variables and $\{\Psi_i\}$ are independent random variables, uniformly distributed in $[0, 2\pi]$.

An alternative spectral representation model, which is equivalent to the one of Eq. (2.184) in the sense that they both converge to the same second order moments as $m \to \infty$, is the following:

$$\hat{X}(t) = \sqrt{2} \sum_{i=0}^{m} \sqrt{2\Phi_{XX}(\omega_i)\Delta\omega} \cos(\omega_i t + \Psi_i) \qquad (2.185)$$

wherein the random variables $\{\Psi_i\}$ are again independent and uniformly distributed in $[0, 2\pi]$. The model in Eq. (2.185) approaches a Gaussian random process if m is sufficiently large due to the central limit theorem [122]. Moreover, the realizations generated by Eq. (2.185) are ergodic in the mean and second moment functions, provided that the contribution of the zero frequency is neglected [122].

Both spectral representations of Eq. (2.184) and Eq. (2.185) yield periodic realizations with period $T_0 = 2\pi/\Delta\omega$ [45]. Therefore, if the domain of interest is [0, T] the incremental frequency $\Delta\omega$ should be chosen such that $T_0 \geq T$. The variance of the truncation error of either spectral representation turns out to be, after basic algebra:

$$\text{Var}\left[X(t) - \hat{X}(t)\right] = \sigma_X^2 - \sum_{i=0}^{m} 2\Phi_{XX}(\omega_i)\Delta\omega \qquad (2.186)$$

Note that Eq. (2.186) is constant with respect to t. Moreover, the second term in Eq. (2.186) is identical to the variance of $\hat{X}(t)$. Therefore, the variance of the error is equal to the difference between the variances of $X(t)$ and $\hat{X}(t)$. Since the variance of the error is always positive, it follows that $\hat{X}(t)$ always under-represents the variance of $X(t)$.

The spectral representation method can also be applied to multi-dimensional weakly homogeneous fields $X(t)$ [123]. As an example, the form in Eq. (2.185) can be extended for two-dimensional fields $X(t)$ as follows [123]:

$$\hat{X}(\mathbf{t}) = \sqrt{2} \sum_{i=0}^{m_1} \sum_{j=0}^{m_2} \sqrt{2\Phi_{XX}(\boldsymbol{\omega}_{ij})\Delta\boldsymbol{\omega}} \cos\left(\boldsymbol{\omega}_{ij}^T \mathbf{t} + \Psi_{ij}\right) \qquad (2.187)$$

wherein $\boldsymbol{\omega}_{ij} = [i\Delta\omega_1, j\Delta\omega_2]^T$, $\Delta\boldsymbol{\omega} = [\omega_{c1}/m_1, \omega_{c2}/m_2]^T$, $\{\Psi_{ij}\}$ are independent random variables uniformly distributed in [0, 2π] and ω_{c1}, ω_{c2} are cut-off frequencies.

The given expressions of the spectral representation method are applicable for stationary (resp. homogeneous) Gaussian processes (resp. fields). The method has been applied for some cases of non-stationary Gaussian processes (e.g. see [31]). Non-Gaussian translation processes have also been approached using the spectral representation of the underlying Gaussian process (e.g. [32], [70], [140]).

The Karhunen-Loève expansion

The Karhunen-Loève (KL) expansion of a random field $X(t)$ is based on the spectral decomposition of its autocovariance function $\Gamma_{XX}(\mathbf{t}_1, \mathbf{t}_2)$ [76]. Let $\{h_i(\mathbf{t})\}$ be a complete set of orthogonal functions, forming a basis in $L^2(\Omega)$, where Ω is the domain of definition of $X(t)$. Since any realization of the

2.5 Discretization of random fields

field belongs in $L^2(\Omega)$, the following expansion of $X(t)$ can be defined, truncated after m terms [144]:

$$\hat{X}(\mathbf{t}) = \mu(\mathbf{t}) + \sum_{i=0}^{m} X_i h_i(\mathbf{t}) \qquad (2.188)$$

where $\{X_i\}$ are zero mean random variables with covariance matrix:

$$\boldsymbol{\Sigma}_{\mathbf{XX}} = \left[\iint_{\Omega\Omega} \Gamma_{XX}(\mathbf{t}_1, \mathbf{t}_2) h_i(\mathbf{t}_1) h_j(\mathbf{t}_2) d\mathbf{t}_1 d\mathbf{t}_2 \right]_{m \times m} \qquad (2.189)$$

It can be shown that if the set $\{h_i(\mathbf{t})\}$ is chosen such that $h_i(\mathbf{t}) = (\lambda_i)^{1/2} \varphi_i(\mathbf{t})$, where $\{\lambda_i, \varphi_i(\mathbf{t})\}$, $(i = 1,\ldots,m)$ are the m largest eigenvalues and corresponding eigenfunctions of the following Fredholm integral equation:

$$\forall i = 1,\ldots \quad \int_{\Omega} \Gamma_{XX}(\mathbf{t}_1, \mathbf{t}_2) \varphi_i(\mathbf{t}_2) d\mathbf{t}_2 = \lambda_i \varphi_i(\mathbf{t}_1) \qquad (2.190)$$

then the mean square truncation error in Eq. (2.188) is minimized [41]. The derived expansion is the KL expansion of $X(\mathbf{t})$, written as:

$$\hat{X}(\mathbf{t}) = \mu(\mathbf{t}) + \sum_{i=0}^{m} X_i \sqrt{\lambda_i} \varphi_i(\mathbf{t}) \qquad (2.191)$$

From Eq. (2.190) it follows that the spectral decomposition of $\Gamma_{XX}(\mathbf{t}_1, \mathbf{t}_2)$ is as follows:

$$\Gamma_{XX}(\mathbf{t}_1, \mathbf{t}_2) = \sum_{i=0}^{\infty} \lambda_i \varphi_i(\mathbf{t}_1) \varphi_i(\mathbf{t}_2) \qquad (2.192)$$

The random variables $\{X_i\}$ in Eq. (2.191) are zero mean and orthonormal, i.e. they are uncorrelated and have unit variance, and are defined as:

$$X_i = \frac{1}{\sqrt{\lambda_i}} \int_{\Omega} [X(\mathbf{t}) - \mu(\mathbf{t})] \varphi_i(\mathbf{t}) d\mathbf{t} \qquad (2.193)$$

Eq. (2.193) implies that if $X(\mathbf{t})$ is Gaussian then the random variables $\{X_i\}$ are also Gaussian and therefore independent. From Eq. (2.191), we can compute the variance of the truncation error, which is shown to be:

$$\text{Var}\left[X(\mathbf{t}) - \hat{X}(\mathbf{t})\right] = \sigma_X^2(\mathbf{t}) - \sum_{i=0}^{m} \lambda_i \varphi_i^2(\mathbf{t}) \tag{2.194}$$

As in Eq. (2.186), the second term in Eq. (2.194) is identical to the variance of $\hat{X}(\mathbf{t})$. Therefore, the KL expansion also under-estimates the point variance of $X(\mathbf{t})$.

The integral eigenvalue problem of Eq. (2.190) can be solved analytically only for rectangular domains and a few autocovariance functions – solutions are given in [41], [127] for one-dimensional homogeneous fields for the triangular and exponential correlation models, extendable to multi-dimensional homogeneous fields in rectangular domains with separable autocorrelation coefficient functions. In any other case, Eq. (2.190) can be solved numerically – see [4] for an overview on the numerical solution of Fredholm integral equations.

A standard procedure for solving Eq. (2.190) in the case where the domain Ω has arbitrary shape is the Galerkin method, using as basis functions the linear shape functions $\{N_i(\mathbf{t})\}$, ($i = 1,\ldots,n$) of a finite element (FE) mesh, such as the one shown in Figure 2.8 ([40], [41]). Note that the set $\{N_i(\mathbf{t})\}$ is defined as in Eq. (2.168), but we will use this notation here due to the conflict with the definition of the eigenfunctions. Also the set $\{N_i(\mathbf{t})\}$ form a complete basis of $L^2(\Omega)$ (if $n \to \infty$), therefore each eigenfunction $\varphi_i(\mathbf{t})$ may be represented as:

$$\varphi_i(\mathbf{t}) \approx \sum_{j=1}^{n} d_i^j N_j(\mathbf{t}) \tag{2.195}$$

where n is the number of nodes of the FE mesh and $\{d_i^j\}$ are unknown coefficients. Substituting Eq. (2.195) to Eq. (2.190) and imposing its satisfaction in the weighted integral sense with arbitrary weighting functions expanded using the basis functions $\{N_i(\mathbf{t})\}$, we obtain the following generalized eigenvalue equation:

$$\Sigma\mathbf{D} = \Lambda\mathbf{M}\mathbf{D} \tag{2.196}$$

where:

2.5 Discretization of random fields

$$\mathbf{M} = \left[\int_\Omega N_i(\mathbf{t}) N_j(\mathbf{t}) d\mathbf{t} \right]_{n \times n} \tag{2.197}$$

$$\mathbf{\Sigma} = \left[\int_\Omega \int_\Omega \Gamma_{XX}(\mathbf{t}_1, \mathbf{t}_2) N_i(\mathbf{t}_1) N_j(\mathbf{t}_2) d\mathbf{t}_1 d\mathbf{t}_2 \right]_{n \times n} \tag{2.198}$$

$$\mathbf{D} = \left[d_i^j \right]_{n \times n} \tag{2.199}$$

$$\mathbf{\Lambda} = \mathrm{diag}[\lambda_i]_{n \times n} \tag{2.200}$$

If instead of the FE shape functions, we use as basis functions in Eq. (2.195) a set $\{h_i(\mathbf{t})\}$ of orthogonal functions in $L^2(\Omega)$, then the derived solution approach is the spectral Galerkin method and Eqs. (2.196)-(2.200) hold for $N_i(\mathbf{t}) = h_i(\mathbf{t})$. A more convenient form of Eq. (2.196) for the FE method can be obtained if the covariance function is projected onto the space spanned by $\{N_i(\mathbf{t})\}$ [66]. Since the functions $\{N_i(\mathbf{t})\}$ are linear interpolating functions, the projection reads:

$$\Gamma_{XX}(\mathbf{t}_1, \mathbf{t}_2) \approx \sum_{i=1}^n \sum_{j=1}^n (\mathbf{\Sigma}_{XX})_{ij} N_i(\mathbf{t}_1) N_j(\mathbf{t}_2) \tag{2.201}$$

where $\mathbf{\Sigma}_{XX}$ is the matrix containing the values of the autocovariance function at the nodal points $\mathbf{t}_i$, i.e.

$$\mathbf{\Sigma}_{XX} = \left[\Gamma_{XX}(\mathbf{t}_i, \mathbf{t}_j) \right]_{n \times n} \tag{2.202}$$

Substituting Eq. (2.201) to Eq. (2.198), we obtain the following generalized eigenvalue equation [66]:

$$\mathbf{M \Sigma_{XX} M D} = \mathbf{\Lambda M D} \tag{2.203}$$

Eqs. (2.196) and (2.203) are discrete and can be solved for $\mathbf{D}$ and $\mathbf{\Lambda}$, using any direct or iterative method. The approximate solution of the Fredholm equation of Eq. (2.203) is equivalent to applying the KL expansion to the field, approximated by the shape function method [66]. Therefore, the minimum variance error of the approximated KL expansion is obtained for

$m = n$ and is identical to the variance error obtained by the shape function method with the same FE mesh.

The KL expansion can be applied for the discretization of Gaussian fields. It was also shown that approximate KL expansions can be constructed for domains with arbitrary shape. In addition, it should be noted that in the case where Ω is rectangular and Eq. (2.190) is solved by the spectral Galerkin method, using the set $\{h_i(\mathbf{t})\}$, $(i = 1,...,n)$ to approximate the eigenfunctions, then the derived KL expansion coincides with the one of Eq. (2.188) if $n = m$ [144]. Moreover, if the field is weakly homogenous, the covariance function is periodic with period equal to the rectangular domain of interest and each eigenfunction is represented by its Fourier series, the KL expansion coincides with the spectral representation of Eq. (2.184) [47].

The expansion optimal linear estimation method

This method [72] is an extension of the optimal linear estimation method described in Section 2.5.1. It is based on the spectral decomposition of the covariance matrix $\mathbf{\Sigma}_{\mathbf{XX}}$ of the nodal random variables $\mathbf{X}$. The random vector $\mathbf{X}$ can be expressed in terms of a vector $\mathbf{U}$ of uncorrelated random variables with zero mean and unit variance, as follows:

$$\mathbf{X} = \mathbf{\mu}_{\mathbf{X}} + \sum_{i=1}^{n} U_i \sqrt{\lambda_i} \mathbf{\varphi}_i \qquad (2.204)$$

where $\{\lambda_i, \mathbf{\varphi}_i\}$, $(i = 1,...,n)$ are the eigenvalues and corresponding eigenvectors of the matrix $\mathbf{\Sigma}_{\mathbf{XX}}$, verifying

$$\mathbf{\Sigma}_{\mathbf{XX}} \mathbf{\varphi}_i = \lambda_i \mathbf{\varphi}_i \qquad (2.205)$$

If the vector $\mathbf{X}$ is Gaussian then $\{U_i\}$ are independent standard normal random variables. A truncation of the sum in Eq. (2.204) after m terms leads to an approximation of the random vector $\mathbf{X}$. Moreover, this approximation corresponds to the optimal low rank approximation of the covariance matrix $\mathbf{\Sigma}_{\mathbf{XX}}$ with respect to the Frobenius norm, due to the Eckart-Young theorem. Substituting Eq. (2.204), truncated after m terms, to

Eq. (2.169) and solving the optimization problem of Eq. (2.170) we get [72]:

$$\hat{X}(\mathbf{t}) = \mu(\mathbf{t}) + \sum_{i=1}^{m} \frac{U_i}{\sqrt{\lambda_i}} \boldsymbol{\varphi}_i^T \boldsymbol{\Sigma}_{X(\mathbf{t})\mathbf{X}} \qquad (2.206)$$

The variance of the truncation error in Eq. (2.206) is:

$$\mathrm{Var}\left[X(\mathbf{t}) - \hat{X}(\mathbf{t})\right] = \sigma_X^2(\mathbf{t}) - \sum_{i=1}^{m} \frac{1}{\lambda_i} \left(\boldsymbol{\varphi}_i^T \boldsymbol{\Sigma}_{X(\mathbf{t})\mathbf{X}}\right)^2 \qquad (2.207)$$

The second term in Eq. (2.207) is identical to the variance of $\hat{X}(\mathbf{t})$. Therefore, the expansion optimal linear estimation under-estimates the point variance of $X(\mathbf{t})$. For a fixed number of nodal random variables $\mathbf{X}$, the minimum variance error is obtained for $m = n$ and is identical to the one given in Eq. (2.175). In principle, the basic idea of this method, i.e. the spectral decomposition of the covariance matrix, can be applied to any point or average discretization method in order to reduce the number of random variables derived from the discretization, while retaining the random components with the largest contribution to the covariance matrix.

2.5.4 Comparison of the discretization methods

This section attempts an assessment of the random field discretization methods presented in Sections 2.5.1-2.5.3. To this end, a one-dimensional homogeneous Gaussian field with zero mean and unit variance, defined over a domain [0,10], is used as test example. Two different autocorrelation coefficient functions are considered, exponential [Eq. (2.116)] and Gaussian [Eq. (2.117)]. The error measure used is the relative point-wise variance error defined in Eq. (2.163).

The point and average discretization methods are assessed first, due to the fact that the random variables derived from the discretization are based on the selection of a SFE mesh for both categories. Figure 2.10 shows the distribution of the error over the domain for the two considered correlation models with a scale of fluctuation $\theta = 5$ and a SFE mesh of four elements with equal size.

58   2 Modeling of uncertainties

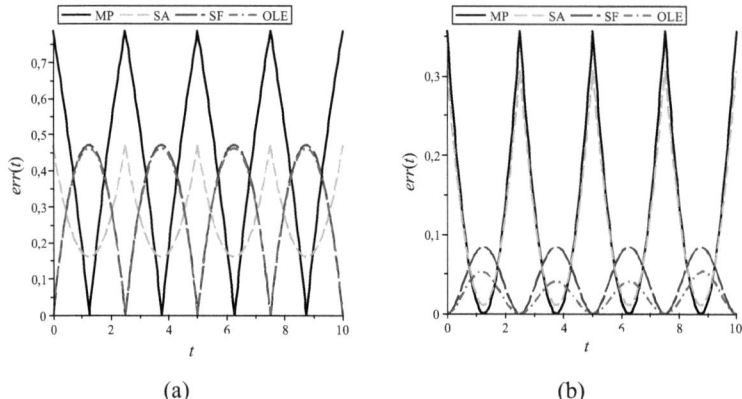

Figure 2.10: Point-wise variance error – comparison of the midpoint method (MP), spatial average method (SA), shape function method (SF) and optimal linear estimation method (OLE). Discretization with 4 elements. (a) exponential (b) Gaussian autocorrelation coefficient function with $\theta = 5$.

As expected, all point discretization methods lead to a zero variance error at the points corresponding to the random variables used in the discretization, i.e. the midpoints of the elements for the midpoint (MP) method and the nodal points for the shape function (SF) and the optimal linear estimation (OLE) methods. The largest error for the MP method and the spatial average (SA) method is obtained at the nodal points, while for the SF and OLE methods at the midpoints of the elements. As also shown in [72], the exponential correlation model, which implies a non-differentiable random field, leads to larger errors compared to the Gaussian model. For the exponential model, the MP method presents the largest overall maximum error, while for the other methods the maximum error is similar. On the other hand, the OLE method clearly outperforms the other methods in the case where the correlation structure is described by the Gaussian model.

In Figure 2.11, the spatial average of the error over the domain is plotted against an increasing number of elements of equal size for all methods, for the two considered correlation models with scales of fluctuation $\theta = 5$ and $\theta = 1$.

2.5 Discretization of random fields

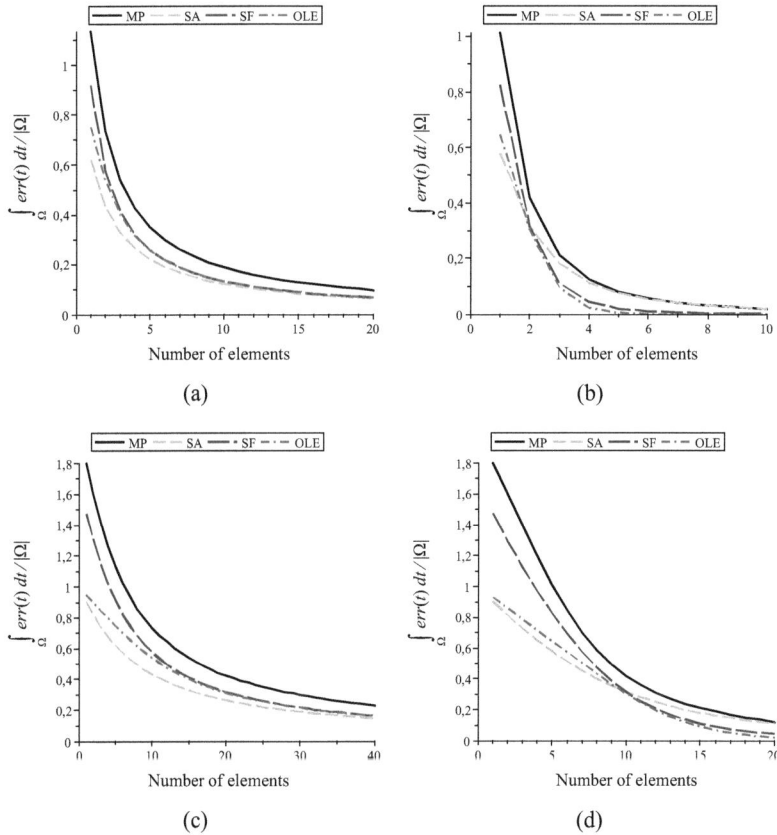

Figure 2.11: Mean variance error against number of elements – comparison of the midpoint method (MP), spatial average method (SA), shape function method (SF) and optimal linear estimation method (OLE). (a), (c) exponential (b), (d) Gaussian autocorrelation coefficient function. (a), (b) $\theta = 5$. (c), (d) $\theta = 1$.

It is shown that the SA method presents the smaller average error for the case of the exponential correlation model at any number of elements, although for a large number of elements the error obtained by the SF and OLE methods is similar. On the other hand, the OLE method presents the fastest convergence in the case of the Gaussian correlation model. In this case, the SA method performs better than the other methods for an element

60   2 Modeling of uncertainties

size greater than the scale of fluctuation, although for smaller element sizes the average error approaches the one obtained by the MP method.

Next, the series expansion methods presented in Section 2.5.3 are compared for the same test example. Figure 2.12 shows the distribution of the relative point-wise variance error, obtained by the KL expansion and the expansion optimal linear estimation (EOLE) method for the two considered correlation models with a scale of fluctuation $\theta = 5$ and 4 number of terms in the expansions. For the KL expansion, we consider the case where the approximate Fredholm eigenvalue problem [Eq. (2.203)] is solved by the FE method (KL-FE) using 20 elements of equal size as well as the case where the eigenvalue problem is solved using the spectral Galerkin method with Legendre polynomials of order 10 (KL-LE). Also 20 elements are used in the EOLE method.

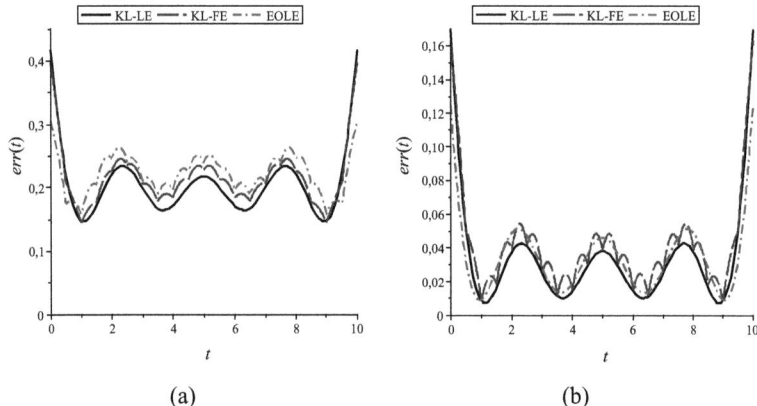

Figure 2.12: Point-wise variance error – comparison of the KL expansion approximated by 10 Legendre polynomials (KL-LE), KL expansion solved by the FE method (KL-FE) and expansion optimal linear estimation method (EOLE). Discretization with 20 elements. Order of expansion: 4. (a) exponential (b) Gaussian autocorrelation coefficient function with $\theta = 5$.

As also shown in [72], the point-wise variance error at the boundaries is larger for the KL than for the EOLE method. The KL-LE performs better than the KL-FE, since the FL-FE solves the integral eigenvalue problem for

2.5 Discretization of random fields

the approximated field by the SF method. Moreover, the eigenfunctions of the considered covariance kernels (as well as for most covariance models that are relevant in the modeling of spatial variability) are in general smooth functions (i.e. they have infinite continuous derivatives) and spectral methods are known to perform better than FE methods in the approximation of smooth functions (e.g. see [4]).

Figure 2.13 shows plots of the spatial average of the error over the domain against the number of terms for all series expansion methods, for the two considered correlation models with scales of fluctuation $\theta = 5$ and $\theta = 1$. It is shown that the KL and EOLE methods perform better than the spectral representation (SR) method for a larger scale of fluctuation (i.e. for strongly correlated random fields). Conversely, for a smaller scale of fluctuation the SR method is more efficient, although the other methods still obtain smaller average errors for a large number of terms in the expansions. This result was also obtained in [128], wherein the KL eigenvalue problem was solved by the wavelet-Galerkin method. In the case where the correlation structure is described by the exponential model, the KL-FE method performs better than the EOLE method, while the reverse holds in the case of the Gaussian model. This can be explained by the fact that the KL-FE method performs the KL expansion of the field approximated by SF method, which is outperformed by the OLE method for the case of the Gaussian correlation model – note that the EOLE method is based on the spectral decomposition of the covariance matrix of the random variables obtained by the OLE method. On the other hand, the KL-LE method performs better than both the KL-FE and EOLE method in all cases. Moreover, it is shown that the Gaussian correlation model presents a faster decay of the average error compared to the exponential model for the EOLE and KL methods. This is due to the fact that for Fredholm eigenvalue problems the smoother the covariance kernel the faster the eigenvalue decay (e.g. see [67], [134]) and therefore the higher the contribution of the larger eigenvalues.

62 2 Modeling of uncertainties

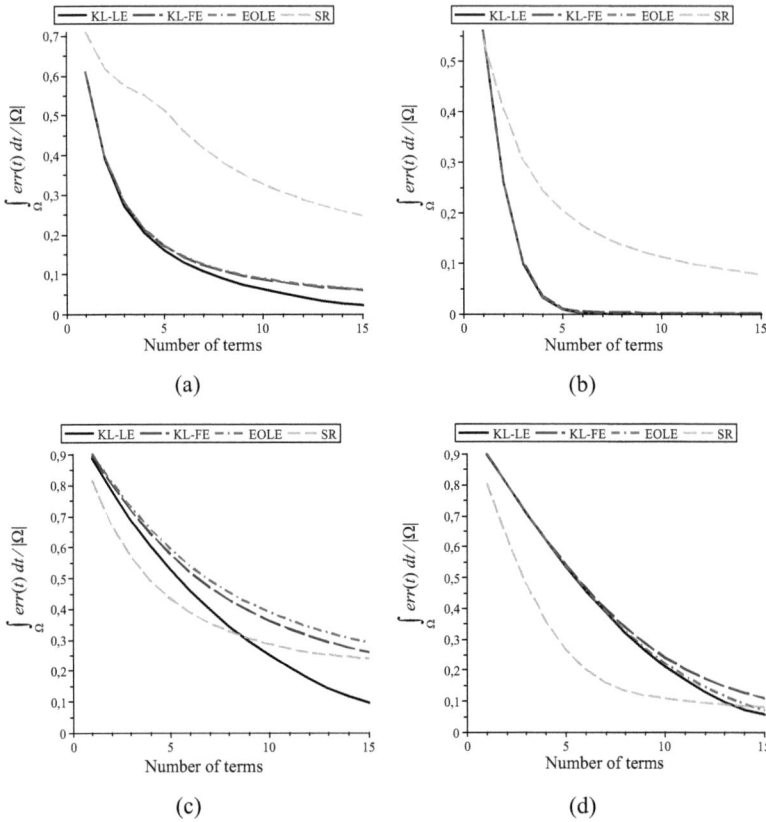

Figure 2.13: Mean variance error against number of terms – comparison of the KL expansion approximated by 20 Legendre polynomials (KL-LE), KL expansion solved by the FE method (KL-FE) and expansion optimal linear estimation method (EOLE). Discretization with 40 elements. (a), (c) exponential (b), (d) Gaussian autocorrelation coefficient function. (a), (b) $\theta = 5$. (c), (d) $\theta = 1$.

2.5.5 Embedded-domain discretization of random fields

In Section 2.5.4, it was shown that the spectral Galerkin method is more efficient in solving the integral eigenvalue problem associated with the KL expansion in one-dimensional domains, as compared to the FE method.

2.5 Discretization of random fields

This method can be easily extended for application to rectangular domains of any number of dimensions, without the need for a generation of a FE mesh. This section investigates the possibility of embedding the actual domain Ω over which the random field is defined in a rectangular volume Ω_r (see Figure 2.14) and using the latter to solve the integral eigenvalue problem of Eq. (2.190), applying the spectral Galerkin method. In this case, the derived eigenvalues λ^r_i and corresponding eigenfunctions $\varphi^r_i(t)$ form a spectral decomposition of the autocovariance function $\Gamma_{xx}(t_1,t_2)$ within the domain Ω_r, which includes the actual domain Ω. Note that the eigenpairs $\{\lambda^r_i, \varphi^r_i(t)\}$ are in general different from those for the actual domain $\{\lambda_i, \varphi_i(t)\}$. Furthermore, the functions $\varphi^r_i(t)$ lose their optimality in Ω, since they are no longer orthogonal in $L^2(\Omega)$. The derived expansion reads:

$$\hat{X}(t) = \mu(t) + \sum_{i=0}^{m} X_i \sqrt{\lambda^r_i}\, \varphi^r_i(t) \tag{2.208}$$

where if the field is Gaussian, the random variables $\{X_i\}$ are independent standard normal random variables. The pair $\{\lambda^r_i, \varphi^r_i(t)\}$, $(i = 1,\ldots,m)$ are the m largest eigenvalues and corresponding eigenfunctions of the following integral equation:

$$\forall i = 1,\ldots \quad \int_{\Omega_r} \Gamma_{xx}(t_1,t_2)\varphi^r_i(t_2)dt_2 = \lambda^r_i \varphi^r_i(t_1) \tag{2.209}$$

In cases for which an analytical solution of Eq. (2.209) exists (homogeneous separable fields with exponential or triangular correlation structure), then its numerical solution is avoided. In the general case, any numerical method can be applied for the solution of Eq. (2.209), although the spectral Galerkin method is suggested due to the smoothness of the functions $\varphi^r_i(t)$. The point-wise variance of the truncation error is identical to the variance error in the rectangular domain Ω_r, i.e.

$$\mathrm{Var}\left[X(t) - \hat{X}(t)\right] = \sigma_X^2 - \sum_{i=0}^{m} \lambda^r_i \left(\varphi^r_i(t)\right)^2 \tag{2.210}$$

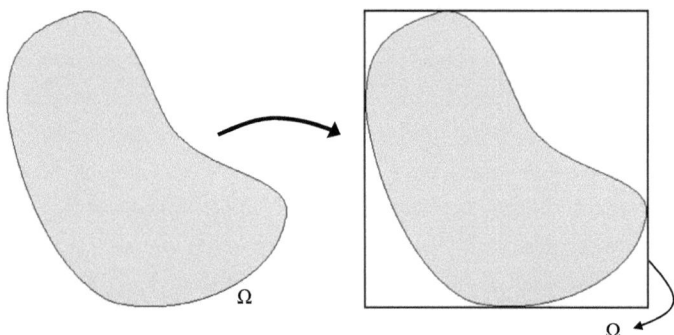

Figure 2.14: Embedded domain discretization of random fields.

The embedded domain KL expansion (KL-ED) can be advantageous for application to complex domains for which the generation of an FE mesh is an involved procedure or to domains which change throughout the computation. An example of the latter case is presented in Section 6.3.

The applicability of the KL-ED approach is demonstrated through a test example of a two-dimensional homogeneous random field with separable correlation structure, defined over four domains of equal volume spanned over a $[-2,2] \times [-2,2]$ rectangular domain, as shown in Figure 2.15. Two different autocorrelation coefficient functions are considered, exponential and Gaussian, with identical correlation lengths is both dimensions. The eigenvalue problem for the KL-ED method is solved using the spectral Galerkin method with Legendre polynomials of order 6. For comparison, the KL-FE approach is applied for the discretization of the field in the actual domain using the FE meshes shown in Figure 2.15. The error measure used is the spatial average of the relative variance error defined in Eq. (2.163).

Figure 2.16 shows plots of the spatial average of the error over the four domains against the number of terms for the two approaches, for the exponential correlation model with one-dimensional scale of fluctuation θ = 4 [since the correlation structure is separable, the correlation parameter η = θ^2 = 16, see Eq. (2.121)]. As expected, the smaller error at any order of expansion is obtained by the KL-FE method, since the approximated

2.5 Discretization of random fields

eigenfunctions are optimal for the actual domain. However, the validity of the KL-ED discretization is verified by its convergent behavior, with an additional eigenfunction required to achieve the same or better level of accuracy when compared to the KL-FE approach.

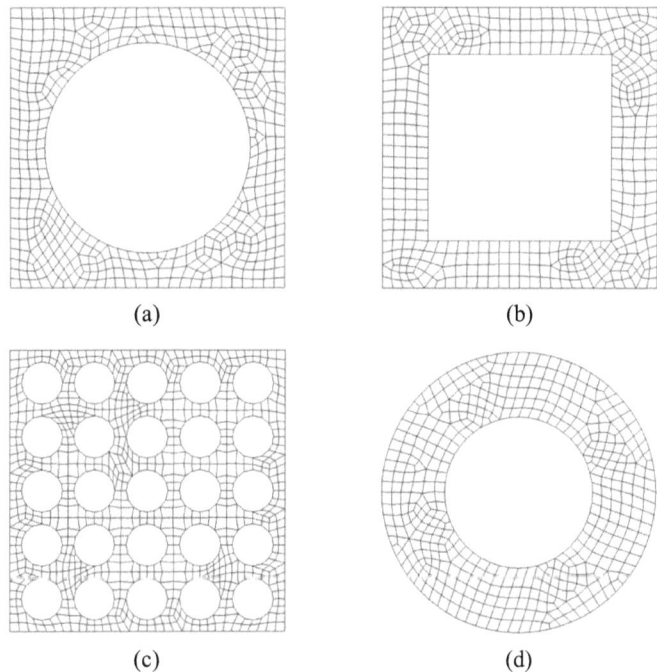

Figure 2.15: Test cases: Four domains with equal volume. FE meshes for the solution of the KL eigenvalue problem.

The performance of the KL-ED method differs for different shapes of the discretization domain. A similar convergence behavior is observed for the two rectangular domains with large holes [Figure 2.16(a) and (b)]; the deviation of the error of the KL-ED from the KL-FE remains constant for the first three terms in the expansion, increases for the next two only to decrease again as the number of terms increases further. This is due to the fact that the fourth and fifth eigenfunctions in the rectangular domain differ considerably from the ones in the actual domain, but this difference is counterbalanced by the information included in the higher eigenfunctions.

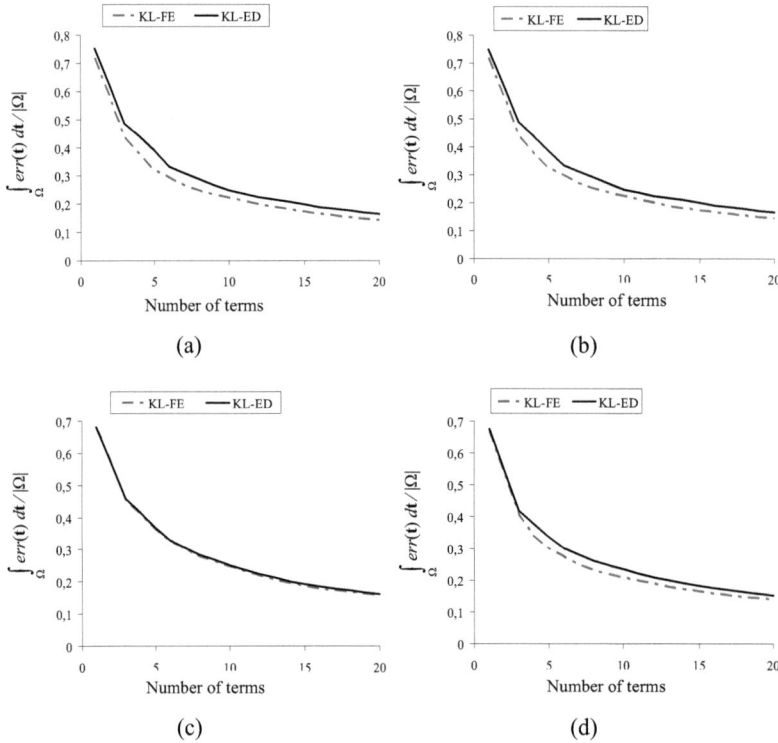

Figure 2.16: Mean variance error against number of terms – comparison of the embedded domain KL expansion approximated by 6th order 2D Legendre polynomials (KL-ED), KL expansion solved by the FE method (KL-FE) at the actual domain. Exponential autocorrelation coefficient function with $\theta = 4$. (a) - (d) as in Figure 2.15.

In the case of the domain with uniform distribution of holes [Figure 2.16(c)], the errors obtained from the two approaches almost coincide at any order of expansion. This can be explained since the first eigenfunctions, which for the chosen scale of fluctuation have the largest contribution to the covariance function, are similar for the actual and

2.5 Discretization of random fields 67

rectangular domains. This is due to the shape of the actual domain presenting a global structure analogous to the fully populated rectangular domain. The fourth case of the disc-shaped domain [Figure 2.16(d)] presents a similar convergence behavior compared to the first two cases [Figure 2.16(a) and (b)], although a very good agreement between the errors obtained by the KL-ED and KL-FE is observed at the first three terms in the expansions.

Figure 2.17 shows comparisons of the errors obtained by the two approaches for the case of a Gaussian autocorrelation coefficient function with one-dimensional scale of fluctuation $\theta = 4$. Comparing Figure 2.17 with Figure 2.16, we observe a similar convergence behavior of the KL-ED for the two correlation models, although its deviation from the KL-FE is smaller for the Gaussian model. Also the Gaussian model implies a differentiable random field which guarantees smaller errors and faster convergence as compared to the non-differentiable exponential model. Therefore, both the KL-ED and KL-FE approaches obtain small average errors with just a few terms in the expansions for the chosen scale of fluctuation.

Figure 2.18 shows plots of the errors obtained for the exponential correlation structure with a smaller scale of fluctuation ($\theta = 2$) as compared to the one used in the results shown in Figure 2.16. The results are shown for the exemplary cases of the domain with circular hole [Figure 2.18(a)] and the domain with uniform distribution of holes [Figure 2.18(b)]. In both cases, the deviation of the errors obtained by the two methods increases with a smaller scale of fluctuation. This is due to the fact that the contribution of the higher eigenfunctions, which are more sensitive to the geometry of the actual domain, becomes larger.

In conclusion, the test examples show that the KL-ED method can be applied for the discretization of random fields in non-rectangular domains. The performance of the method is sensitive to the shape of the actual domain and the chosen correlation model; a more efficient performance should be expected at geometries with a uniform distribution of holes, such as porous domains; also the method converges faster when the smoothness of the correlation model increases. Finally, the performance of the method becomes poorer for small scales of fluctuation. In any case, the error measure given in Eq. (2.210) can be used for a decision on whether the method is applicable to specific problems.

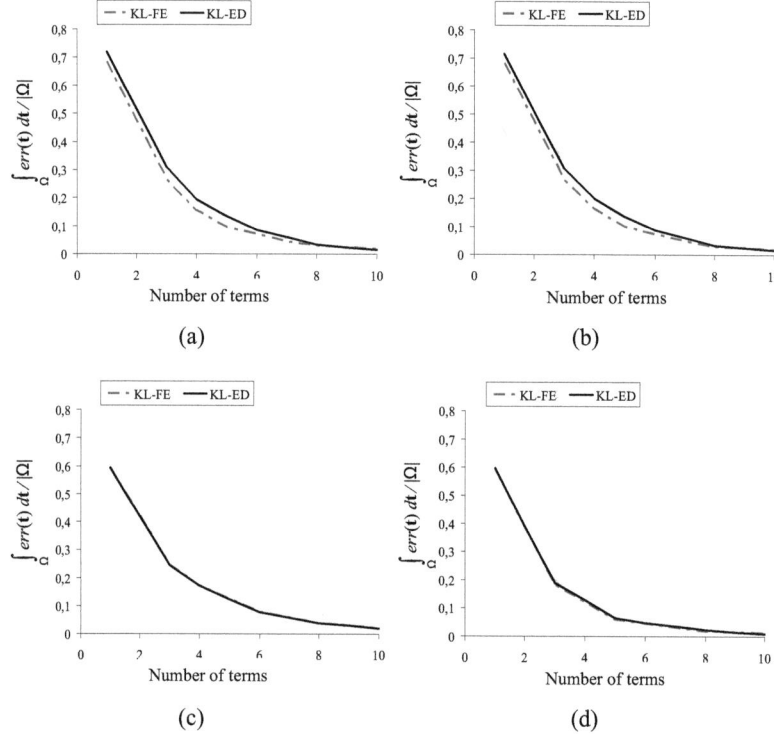

Figure 2.17: Mean variance error against number of terms – comparison of the embedded domain KL expansion approximated by 6th order 2D Legendre polynomials (KL-ED), KL expansion solved by the FE method (KL-FE) at the actual domain. Gaussian autocorrelation coefficient function with $\theta = 4$. (a) - (d) as in Figure 2.15.

2.5 Discretization of random fields

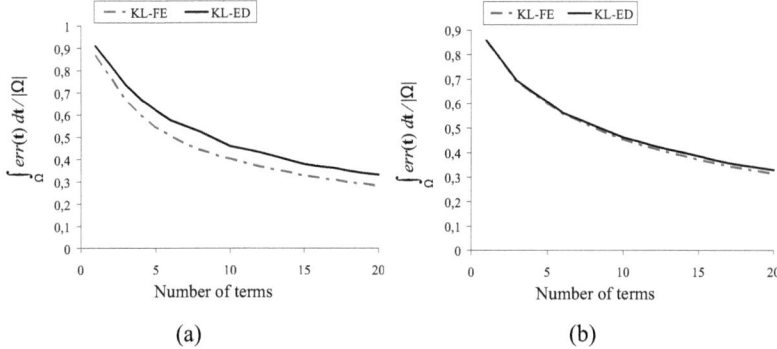

Figure 2.18: Mean variance error against number of terms – comparison of the embedded domain KL expansion approximated by 6th order 2D Legendre polynomials (KL-ED), KL expansion solved by the FE method (KL-FE) at the actual domain. Exponential autocorrelation coefficient function with $\theta = 2$. (a) as in Figure 2.15(a). (b) as in Figure 2.15(c).

3 Fundamental concepts of reliability analysis

This chapter discusses the existing design concepts and provides the fundamental ideas and basic definitions used in the reliability analysis of structural systems. Furthermore, the important concepts of the probability of failure and reliability index are introduced.

3.1 Evolution of design concepts

For a long time, the structural design process was almost entirely based on empirical knowledge, which had primarily been gained by trial and error. Along with the evolution of the theory of structures, material science and computational possibilities, the first design concepts were established. In the beginning of this process, simple instructions and guidelines documented the state of knowledge. Successively, a comprehensive system of technical directives and codes was developed. The underlying safety concept was and is continuously adapted to the attained practical and experimental experience, the increased theoretical knowledge and the computational possibilities.

Currently adopted design concepts can be classified into deterministic, semi-probabilistic and full-probabilistic approaches. Differences can

3 Fundamental concepts of reliability analysis

readily be contrasted, when considering the design goal 'structural safety' (ultimate limit-state). The common background of all approaches is the control of the vast amount of inherent uncertainties. These uncertainties, as discussed in Chapter 1, are induced by variation of material properties, by the limited predictability of loading or by the construction process itself. Hence, the objective can be identified as limiting the probability of failure to an acceptable level.

Let us assume that the ultimate limit-state criterion can be expressed in terms of the resistance (capacity) R and the effect of the loading (demand) S. The values of R and S are assumed to be uncertain and described by probability density functions $f_R(r)$, $f_S(s)$ respectively. The safety limit-state will be violated if:

$$R - S \leq 0 \qquad (3.1)$$

Of course, both the resistance and the loading can be functions of time. In this case, the limit-state reads:

$$R(t) - S(t) \leq 0 \qquad \forall t \in [0, T] \qquad (3.2)$$

where T represents the life expectancy of the structure.

The deterministic design approach adopts the simplest safety definition:

$$R/\gamma \geq S \qquad (3.3)$$

where γ is an empirically determined global safety factor ($\gamma > 1$) that is meant to incorporate all uncertainties. This approach does not account for the distributions of R, S and ignores their relative influences on the structural performance [see Figure 2.11(a)].

A more refined version is represented by partial safety factors:

$$R_k/\gamma_R \geq S_k \cdot \gamma_S \qquad (3.4)$$

In this case, uncertainties on action and resistance side are accounted for, separately. They are reflected by the load factor $\gamma_S (>1)$ and the resistance factor $\gamma_R (>1)$ [Figure 2.11(b)]. The values R_k and S_k are characteristic values usually chosen as the 5% and 95% exceedance values of the

3.1 Evolution of design concepts

probability distribution of the respective parameter. Computation remains purely deterministic but since the partial safety factors are calibrated to match a predefined probability of failure level P_d (=acceptable probability of failure), the approach is classified as semi-probabilistic.

The full probabilistic approach finally abandons the definition of (partial) safety factors; it simply imposes the constraint on the probability of failure P_f directly [Figure 2.11(c)]:

$$P_f = P(R - S \leq 0) < P_d \qquad (3.5)$$

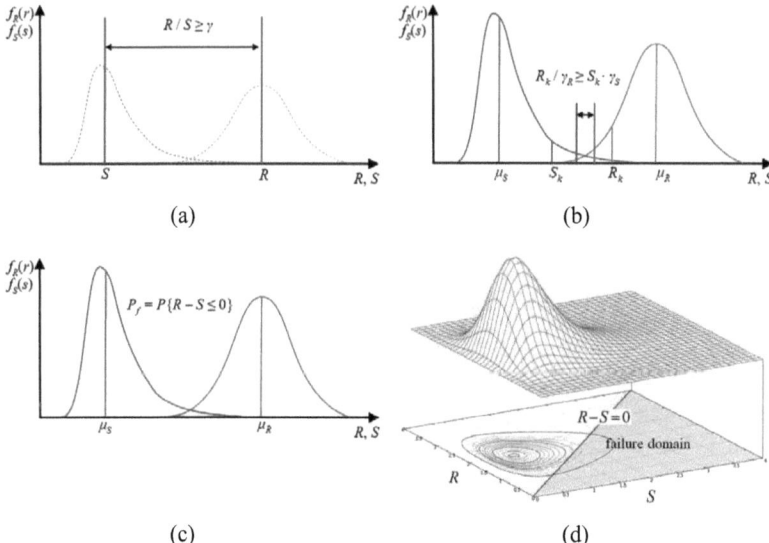

Figure 3.1: Illustration of safety concepts. (a) Deterministic safety concept. (b) Partial safety factors. (c), (d) Probabilistic safety concept.

This most general approach ultimately leaves the domain of deterministic calculation. Also, as will be illustrated further, the concept can be employed for serviceability assessment in a straight-forward manner; in fact, any criterion that is considered relevant for successful performance of the structure can be adopted. Moreover, this concept allows for the

74   3 Fundamental concepts of reliability analysis

definition of limit-states in terms of several random variables, possibly correlated, which may account for all different sources of uncertainty. Next the evaluation of the probability in Eq. (3.5) is discussed. Section 3.3 generalizes the definition of the probability of failure for arbitrary limit-state criteria and number of random variables.

3.2   The elementary reliability problem

Consider the reliability problem of Eq. (3.5), wherein the limit-state criterion is defined in terms of the load effect S and the resistance R, each of which is described by a known probability density function $f_S(s)$, $f_R(r)$ respectively. The value of S may be expressed in terms of the applied load, say Q, through a structural analysis. For convenience, we consider here the failure of some specific structural component. In this case, the problem is termed component reliability problem.

The probability of failure P_f of the structural component is as follows:

$$P_f = P(R - S \leq 0)$$
$$= P\left[g(R,S) \leq 0\right] \tag{3.6}$$

where $g(.)$ is termed 'limit-state function' with negative values defining the failure scenario. The random variables R, S have a joint probability density function denoted by $f_{R,S}(r, s)$. In Figure 2.11(d), the joint PDF is plotted as well as the hatched failure domain, so that the probability of failure becomes:

$$P_f = P(R - S \leq 0) = \int\int_{r-s\leq 0} f_{R,S}(r,s) \, dr \, ds \tag{3.7}$$

If the two variables are statistically independent, then the joint PDF is expressed as:

$$f_{R,S}(r,s) = f_R(r) f_S(s) \tag{3.8}$$

It should be noted that Figure 2.11(a)-(c) have already assumed independence of R and S. In this case, Eq. (3.7) becomes:

3.2 The elementary reliability problem

$$P_f = \int_{-\infty}^{\infty} \int_{-\infty}^{r \leq s} f_R(r) f_S(s) \, dr \, ds \tag{3.9}$$

Considering that

$$F_R(x) = \int_{-\infty}^{x} f_R(z) \, dz \tag{3.10}$$

where $F_R(x)$ is the cumulative distribution function of R, we can write Eq. (3.9) as:

$$P_f = \int_{-\infty}^{\infty} F_R(x) f_S(x) \, dx \tag{3.11}$$

Eq. (3.11) expresses the failure probability as an infinite sum of the infinitesimal areas $f_S(x)dx$ multiplied by the integrals of $f_R(z)$ in the limits $[-\infty, x]$. This can be explained as the infinite sum of all probabilities that the resistance is smaller than a value of the load effect, over all possible values of the load effect. An alternative expression for Eq. (3.11) is:

$$P_f = \int_{-\infty}^{\infty} \left[1 - F_S(x)\right] f_R(x) \, dx \tag{3.12}$$

which is the infinite sum of all probabilities that the load effect exceeds a value of the resistance, over all possible values of resistance.

In some special cases, the integral of Eq. (3.7) can be computed analytically. One example is when the variables R, S are normal random variables with means μ_R, μ_S and standard deviations σ_R, σ_S respectively. The random variable $Z = R - S$ will then be a normal random variable with mean and standard deviation as follows:

$$\mu_Z = \mu_R - \mu_S \tag{3.13}$$

$$\sigma_Z = \sqrt{\sigma_R^2 + \sigma_S^2 - 2\text{Cov}[R,S]} \tag{3.14}$$

for the general case where R, S are correlated. Eq. (3.7) then takes the following form:

76   3 Fundamental concepts of reliability analysis

$$P_f = P(R - S \leq 0) = P(Z \leq 0) = F_Z(0) = \Phi\left(\frac{-\mu_Z}{\sigma_Z}\right) = \Phi(-\beta) \qquad (3.15)$$

where $\Phi(.)$ denotes the standard normal CDF. The quantity $\beta = \mu_Z / \sigma_Z$ indicates the number of standard deviations between the mean value μ_Z and the limit state $Z = 0$ (Figure 3.2). This quantity, introduced by Cornell [16], is defined as the reliability index. However, this definition of the reliability index is not generally valid, even for the elementary R,S-case, due to its lack of invariance for different equivalent definitions of the limit-state criterion (see Section 4.3 for an example). Alternative definitions of the reliability index are discussed in Chapter 4, while Section 3.3.2 gives a rigorous generally valid definition of this reliability measure.

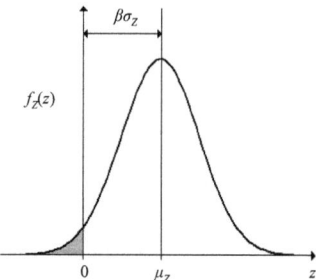

Figure 3.2:   Distribution of $Z = R - S$ and reliability index β.

3.3   The generalized reliability problem

In many cases, the definition of the probability of failure given in Eq. (3.6) is not sufficient, since it may not be possible to reduce the design condition to a simple R versus S relation. In a more general context, there may be many different parameters that are expected to present an uncertain behavior. Typical examples are dimensions, loads, material properties as well as any other variable that is employed in structural analysis and design procedures. The variables which define the behavior and safety of a structure are termed 'basic' variables.

3.3 The generalized reliability problem 77

The basic variables can generally be dependent and are characterized by a joint probability density function. The joint PDF can be estimated by first estimating all conditional PDF's and then applying the multiplication rule (see Section 2.3.1). Alternatively, the joint PDF can be approximated by the Nataf distribution which reduces to the estimation of the marginal PDF's of all basic variables and the correlation matrix (see Section 2.3.4). In this case, the dependence of the random variables is completely described by the correlation matrix.

In the case where basic variables are expected to present some spatial or time variability, then a complete definition would require the joint PDF for any number of random variables corresponding to specific points in space or time. However, this is not feasible in practice and the definition is given using second-moment information and the marginal distribution (see Section 2.4.6). Therefore, the estimation of the second-moment functions, including the autocorrelation structure of the random field or process, is additionally required. Furthermore, the random field or process can be reduced to a finite number of random variables, applying any of the discretization methods described in Section 2.5.

The choice of the probability distributions assigned to the basic variables should depend on available data and engineering judgment. In some cases, the central limit theorem may be applied, when a basic variable consists of a superposition of many different variables. Also, in several cases physical reasoning can be used, such as in the case of extreme value distributions. For example, the maximum wind velocity over a given time span can be modeled by the Gumbel distribution, assuming that the daily wind velocity is normally distributed. Moreover, many physical variables require lower or upper bounds. For example, the Young's modulus of a material can only take positive values. In this case, the lognormal distribution can be used.

The parameters of the distributions can be estimated from available data applying any statistical estimation method, such as the method of moments, the maximum likelihood method and the Bayesian parameter estimation (e.g. see [1]). Similar method can also be used for the estimation of the autocorrelation structure of a random field or process, when an

78   3 Fundamental concepts of reliability analysis

analytical correlation model is used, such as the ones given in Section 2.4.2.

Let the vector $\mathbf{X} = [X_1, X_2, ..., X_n]^T$ represent all the basic random variables involved in the problem. Also, let $f_\mathbf{X}(\mathbf{x})$ be the already established joint PDF of **X**. Any criterion that is considered as relevant for a successful performance of the structure can be expressed by means of a corresponding limit-state function $g(\mathbf{X})$, defined in terms of the vector **X**. The limit-state function $g(\mathbf{X})$ is defined by convention, such that failure of successful performance occurs when $g(\mathbf{X}) \leq 0$. Conversely, satisfactory (safe) performance is guaranteed when $g(\mathbf{X}) > 0$. Therefore, the limit-state equation $g(\mathbf{X}) = 0$ defines the boundary between the safe and unsafe domain in the n-dimensional basic random variable space.

3.3.1 Generalization of the probability of failure

Using the generalized definition of the limit-state function $g(\mathbf{X})$, we can obtain a generalization of Eq. (3.7) as follows:

$$P_f = P[g(\mathbf{X}) \leq 0] = \int \cdots \int_{g(\mathbf{X}) \leq 0} f_\mathbf{X}(\mathbf{x}) dx_1 \ldots dx_n = \int_{g(\mathbf{X}) \leq 0} f_\mathbf{X}(\mathbf{x}) d\mathbf{x} \quad (3.16)$$

where $d\mathbf{x} = dx_1 \ldots dx_n$. The function $f_\mathbf{X}(\mathbf{x})$ is the joint PDF of the basic variables **X** and the region of integration $g(\mathbf{X}) \leq 0$ denotes the failure domain.

In the general case, the integration in Eq. (3.16) cannot be performed analytically, except from some special cases with limited practical interest. Numerical integration can be applied using any simple numerical procedure or quadrature formula. These methods would typically require a truncation of the integration domain using for example an n-dimensional hypercube or hypersphere. However, numerical integration methods have limited applicability due to the geometric increase of the computational cost as the number of random variables increases (the so-called 'curse of dimensionality') – note that each evaluation of the limit-state function may require a time-consuming finite element calculation.

3.3 The generalized reliability problem

In the case where the limit-state function $g(\mathbf{X})$ is expressed as a linear combination of the basic variables X_i ($i = 1,...,n$), and the vector $\mathbf{X}$ is jointly Gaussian, then the variable $Z = g(\mathbf{X})$ will be normally distributed, while its mean μ_Z and standard deviation σ_Z can be computed in a straightforward manner. In this case, the integration in Eq. (3.16) is avoided and the probability of failure can be computed according to Eq. (3.15). Based on this observation, a number of methods that transform the original problem to an equivalent problem in a Gaussian space and perform a first- or second-order approximation of the limit-state function in this space have emerged during the past decades. These approximation methods consist of the first- and second-order reliability methods and will be discussed in detail in Chapter 4.

An alternative integration technique for the evaluation of the integral in Eq. (3.16) is the so-called Monte Carlo integration method. This method belongs to the category of simulation methods, which are based on random sampling of the basic random variable space. These methods are asymptotically exact in the sense that an infinite number of samples would theoretically lead to the exact evaluation of the probability of failure. The crude Monte Carlo method overcomes the 'curse of dimensionality' since its computational cost does not depend on the number of basic random variables. However, the required number of samples is inversely proportional to the value of the probability of failure. Therefore the required number of samples increases dramatically for small values of the expected failure probability, which are typically desired in structural engineering applications. Several simulation methods that overcome this problem have been proposed, such as importance sampling, directional simulation and subset simulation. These methods will be thoroughly discussed in Chapter 4.

3.3.2 Reliability measures

A reliability measure (safety measure) can be chosen as any decreasing function of the probability of failure P_f. A straightforward reliability measure is the probability of successful performance, defined as the probability of the complement of the failure event:

$$P_s = 1 - P_f \qquad (3.17)$$

An equivalent reliability measure is the generalized reliability index [33], defined as:

$$\beta = -\Phi^{-1}(P_f) \qquad (3.18)$$

where $\Phi^{-1}(.)$ is the inverse of the standard normal cumulative distribution function. The definition in Eq. (3.18) is motivated by the relation obtained when inverting Eq. (3.15). However, the reliability index in Eq. (3.18) is not depending on the procedure followed for the evaluation of the probability of failure. It also assumes that an exact (invariant) value of the reliability index exists and depends on the exact value of the probability of failure. Therefore, this definition should not be confused with the definition of the reliability index used in the first- or second-order approximation methods (refer to Chapter 4).

3.4 The system reliability problem

The definition of the probability of failure in Eq. (3.16) is based on a limit-state function $g(\mathbf{X})$, representing the failure condition of one component of the structure, i.e. one failure mode. Consider now the case where a structure has m significant modes of failure, which can be expressed by m corresponding limit-state functions $\{g_i(\mathbf{X}), i = 1,\ldots,m\}$, defined in terms of the vector of basic random variables $\mathbf{X}$. Substituting each $g_i(\mathbf{X})$ for $g(\mathbf{X})$ in Eq. (3.16) will result in m component reliability problems. In reliability analysis of systems, we are interested in the probability of failure of a certain combination of failure events.

A first very important step in the system reliability analysis is the identification of all possible modes of failure of a system and the definition of the combination of modes that describe the failure criterion of the system. To this end, several approaches have been proposed, e.g. the failure

3.4 The system reliability problem

mode approach, and usually result in the construction of fault or event trees (e.g. see [129]).

Denoting the failure event associated with each component limit-state function $g_i(\mathbf{X})$ by F_i, we can define two distinct system reliability problems; the series and the parallel system reliability problem. The series reliability problem is defined by the union of the events $\{F_i, i = 1,\ldots,m\}$. The probability of failure of the system $P_{f,ser}$ is expressed as:

$$P_{f,ser} = P\left(\bigcup_{i=1}^{m} F_i\right) \tag{3.19}$$

The parallel reliability problem is defined by the intersection of the events $\{F_i, i = 1,\ldots,m\}$ with corresponding system probability of failure:

$$P_{f,par} = P\left(\bigcap_{i=1}^{m} F_i\right) \tag{3.20}$$

Consider now K index sets $c_k \subseteq \{1,\ldots,m\}$, defining sub-systems, corresponding to parallel system reliability problems. Then we can define the general system reliability problem as follows:

$$P_{f,sys} = P\left(\bigcup_{k=1}^{K}\left(\bigcap_{i \in c_k} F_i\right)\right) \tag{3.21}$$

From Eq. (3.21) we can retrieve the component reliability problem by setting $m = 1$, the parallel system reliability problem by setting $K = 1$ and the series system reliability problem by defining each set c_k by a single index.

In some cases, it may be convenient to define the general reliability problem of Eq. (3.21) using one equivalent limit-state function, which reads:

$$g(\mathbf{x}) = \min_{1 \leq k \leq K}\left[\max_{i \in c_k} g_i(\mathbf{x})\right] \tag{3.22}$$

The equivalent limit-state functions for the component, parallel and series system reliability problems can be obtained from Eq. (3.22) in the same manner as the corresponding reliability problems were obtained from Eq. (3.21). The probability of failure of any system or component reliability problem can be obtained by substituting Eq. (3.22) to Eq. (3.16), giving:

$$P_{f,sys} = P\left\{\min_{1 \leq k \leq K}\left[\max_{i \in c_k} g_i(\mathbf{x})\right] \leq 0\right\} = \int_{\min_{1 \leq k \leq K}\left[\max_{i \in c_k} g_i(\mathbf{x})\right] \leq 0} f_\mathbf{X}(\mathbf{x}) \, d\mathbf{x} \qquad (3.23)$$

The resulting expression of Eq. (3.23) is suitable for application of simulation methods. However, Eq. (3.23) cannot be solved by the first- or second-order approximation methods. The application of these methods for the solution of system reliability problems will be discussed in Section 4.6.

4 Finite element reliability assessment

This chapter discusses a series of reliability methods that have been combined with a deterministic finite element program in a tool intended for finite element reliability analysis of structural systems. First, an overview of the program framework is given. Then the implemented methods are discussed in detail.

4.1 Introductory comments

An important byproduct of this work is a FORTRAN code designed for reliability analysis of structural systems and coupled to the commercial finite element (FE) program SOFiSTiK [126]. A trial version of the code has been fully integrated in the SOFiSTiK program as an auxiliary tool for reliability analysis, named RELY [96]. RELY can be viewed as a stand-alone program that exchanges information with SOFiSTiK via CDB, which is the SOFiSTiK database.

The input commands and parameters are read from an input file. These include the identification of the random parameters, their probabilistic information (i.e. distributions and correlations), information on the applied reliability methods and the definition of the relevant limit-state functions.

4 Finite element reliability assessment

The limit-state functions can be either of analytical form or products of an FE calculation. In the latter case, the RELY input file is linked to a standard SOFiSTiK input file that defines the details for the finite element analysis. The program stores the basic variables and their probabilistic information in CDB and subsequently performs the required reliability analysis, during which the limit-state functions are evaluated when needed by updating the realizations of the basic variables in CDB and running the corresponding SOFiSTiK modules. Then, the results (e.g. probability of failure, reliability index, sensitivity indices) are printed out in standard SOFiSTiK report files. A schematic diagram of the framework of RELY is shown in Figure 4.1.

It should be noted that the majority of the computational time for the reliability analysis of real-world industrial problems with RELY is related to the evaluations of the limit-state functions. This is due to the fact that for such problems, a considerable computational time is expected for the single finite element calculation. Therefore, we should point out the need for a compromise between accuracy and efficiency in the reliability analysis algorithm, where by efficiency the limiting of the number of evaluations of the limit-state functions is understood. The methods described in the following sections aim at such a compromise.

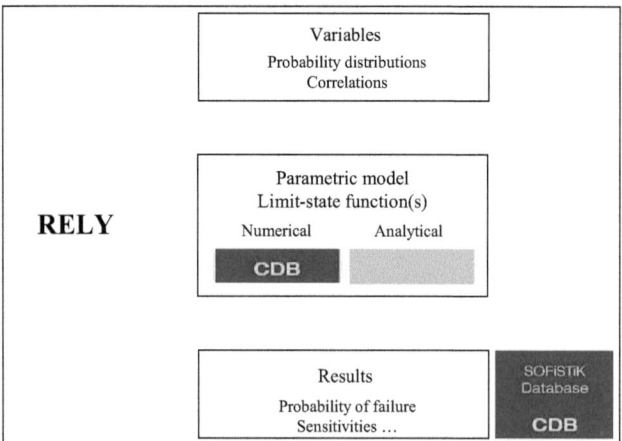

Figure 4.1: Framework of the program RELY

4.2 Isoprobabilistic transformation

The implementation of the methods described in the following sections has been done in the so-called equivalent standard normal space, denoted here by **U**. This is the space of independent standard normal random variables that is derived from an isoprobabilistic transformation **T**(.) of the n-dimensional space of basic random variables **X**, i.e. **U** = **T**(**X**). The generalized reliability problem defined in Eq. (3.16) can be formulated in the **U**-space as:

$$P_f = P[G(\mathbf{U}) \leq 0] = \int_{G(\mathbf{U}) \leq 0} \varphi_n(\mathbf{u}) \, d\mathbf{u} \qquad (4.1)$$

wherein $G(\mathbf{U}) = g[\mathbf{T}^{-1}(\mathbf{U})]$ and $\varphi_n(.)$ is the n-dimensional standard normal PDF.

This section presents the utilized transformation method based on the assumption of the Nataf distribution and briefly discusses an additional approach.

Marginal transformation

In the case where the random components of **X** are statistically independent with strictly increasing continuous marginal distribution functions $\{F_{X_i}(x_i), i = 1,...,n\}$, then the following isoprobabilistic transformation can be defined:

$$Z_i = \Phi^{-1}\left[F_{X_i}(x_i)\right] \qquad \forall i : 1 \leq i \leq n \qquad (4.2)$$

where $\Phi^{-1}(.)$ is the inverse of the standard normal cumulative distribution function. If the components of **X** are not statistically independent then the random variables derived from the marginal transformation of Eq. (4.2) will be dependent. If the probabilistic information of the vector **X** comes in terms of the marginal distributions and the correlation matrix $\Sigma_{\mathbf{XX}}$, then we can model the joint distribution of **X** using the Nataf model (see Section 2.3.4). In this case, the random vector **Z** derived from Eq. (4.2) will be a jointly Gaussian vector of standard normal random variables with correlation coefficient matrix **R**. The elements ρ_{ij} of **R** can be computed from the following integral equation [see also Eq. (2.93)]:

$$\rho_{X_i,X_j} = \int_{-\infty}^{\infty}\int_{-\infty}^{\infty} \left(\frac{x_i - \mu_{X_i}}{\sigma_{X_i}}\right)\left(\frac{x_j - \mu_{X_j}}{\sigma_{X_j}}\right) \varphi_2(z_i, z_j, \rho_{ij}) dz_i dz_j \qquad (4.3)$$

where ρ_{X_i,X_j} are the elements of the correlation coefficient matrix of **X**, μ_{X_i}, σ_{X_i} are the mean and standard deviation of the each component of **X** and $\varphi_2(.)$ is the bivariate Gaussian probability density function. For the solution of Eq. (4.3), the empirical formulae given in [25], [73] for several combinations of distribution types are applied. If the correlation coefficient matrix **R** derived from Eq. (4.3) is positive definite, we can perform its Cholesky decomposition giving:

$$\mathbf{R} = \mathbf{A}\mathbf{A}^T \qquad (4.4)$$

Then we can transform the vector **Z** to an independent standard Gaussian vector **U** as follows:

$$\mathbf{U} = \mathbf{A}^{-1}\mathbf{Z} \qquad (4.5)$$

Eq. (4.5) can be easily verified if we apply Eqs. (2.88) and (2.89) to Eq. (2.85), taking into account that the Jacobian of the transformation reads:

$$\mathbf{J}_{u,z} = \mathbf{A}^{-1} \qquad (4.6)$$

We can then compute the Jacobian of the transformation from the basic random variable space **X** to the equivalent independent standard normal space **U**, by applying the chain rule, which gives:

$$\mathbf{J}_{u,x} = \mathbf{A}^{-1} \text{diag}\left[\frac{f_{X_i}(x_i)}{\varphi(u_i)}\right]_{n\times n} \qquad (4.7)$$

where $\{f_{X_i}(x_i), i=1,\ldots,n\}$ are the marginal probability distribution functions of the components of **X** and $\varphi(.)$ is the standard normal PDF.

Rosenblatt transformation

In the case where a complete description of the random vector **X** is given in terms of the joint distribution function $F_\mathbf{X}(\mathbf{x})$, then the following isoprobabilistic transformation to an equivalent independent standard normal space **U** can be defined ([57], [111]):

$$\begin{aligned} U_1 &= \Phi^{-1}\left[F_{X_1}(x_1)\right] \\ U_2 &= \Phi^{-1}\left[F_{X_2|X_1}(x_2 \mid x_1)\right] \\ &\vdots \\ U_n &= \Phi^{-1}\left[F_{X_n|X_1,\ldots,X_{n-1}}(x_n \mid x_1,\ldots,x_{n-1})\right] \end{aligned} \qquad (4.8)$$

The conditional distribution functions in Eq. (4.8) can be computed by integrating the conditional PDF's derived from Eq. (2.69). The validity of the transformation in Eq. (4.8) can be verified by applying Eq. (2.85), expressing the joint PDF of **X** in terms of the conditional PDF's applying the multiplication rule of Eq. (2.70) and taking into account that the Jacobian of the transformation reads:

$$\mathbf{J}_{u,x} = \operatorname{diag}\left[\frac{f_{X_1}(x_1)}{\varphi(u_1)}, \frac{f_{X_2|X_1}(x_2 \mid x_1)}{\varphi(u_2)}, \ldots, \frac{f_{X_n|X_1,\ldots,X_{n-1}}(x_n \mid x_1,\ldots,x_{n-1})}{\varphi(u_n)}\right]_{n \times n} \qquad (4.9)$$

It should be noted that the Rosenblatt transformation is not invariant when changing the order of the random variables in the construction of the conditional distributions [35].

4.3 The first order reliability method

As also discussed in Section 3.3.1, in the case where the limit-state function $g(\mathbf{X})$ is a linear function of jointly normal random variables $\{X_i \; (i = 1,\ldots,n)\}$, the probability of failure can be computed by:

$$P_f = \Phi(-\beta) \qquad (4.10)$$

where β is the reliability index defined in terms of the statistics of the random variable $Z = g(\mathbf{X})$, as follows:

$$\beta = \frac{\mu_Z}{\sigma_Z} \qquad (4.11)$$

However, as already pointed out in Section 3.2, the definition of Eq. (4.11) is not invariant for equivalent definitions of the limit-state function. This can be explained by the following example. Consider two equivalent limit-state criteria defined by the following limit-state functions:

$$g_1(\mathbf{X}) = X_1 - X_2 \qquad (4.12)$$

$$g_2(\mathbf{X}) = \frac{X_1}{X_2} - 1 \qquad (4.13)$$

Both limit-state criteria, i.e. $g_1(\mathbf{X}) \leq 0$ and $g_2(\mathbf{X}) \leq 0$, define the same failure domain in the basic random variable space. Assuming that X_1, X_2 are statistically independent, the reliability index for $g_1(\mathbf{X})$ can be computed applying Eq. (4.11) for $Z_1 = g_1(\mathbf{X})$:

$$\beta_1 = \frac{\mu_{X_1} - \mu_{X_2}}{\sqrt{\sigma_{X_1}^2 + \sigma_{X_2}^2}} \qquad (4.14)$$

However, $g_2(\mathbf{X})$ is not a linear function of $\mathbf{X}$. Moreover, if $\mathbf{X}$ consists of independent Gaussian components then $Z_2 = g_2(\mathbf{X})$ will be a Cauchy variable, which does not have moments of any order. This problem can be solved if we use the linearization of $g_2(\mathbf{X})$ at some point, e.g. at $\left[\mu_{X_1}, \mu_{X_2}\right]$. In this case, the approximate reliability index of $g_2(\mathbf{X})$ is derived after straightforward calculations as:

$$\beta_2 = \frac{\mu_{X_1} - 1}{\sqrt{\sigma_{X_1}^2 + \frac{\mu_{X_1}^2}{\mu_{X_2}^2}\sigma_{X_2}^2}} \qquad (4.15)$$

which differs significantly from β_1. Note that this approach of computing an approximate reliability index in terms of the moments of the linearized limit-state function is called first order second moment method (e.g. see [1], [83]).

An invariant definition of the reliability index can be formulated in the equivalent standard normal space **U**. In the U-space, every projection of **U** on an arbitrary line passing through the origin is a Gaussian random variable with zero mean value and unit standard deviation. Then a consistent definition of the reliability index can be formulated as the distance from the origin to the limit-state surface $G(\mathbf{U}) = 0$ [53], i.e.

$$\beta = \min\left\{\sqrt{\mathbf{u}^T\mathbf{u}} \,\Big|\, G(\mathbf{u}) = 0\right\} \qquad (4.16)$$

If the limit-state function is linear in the U-space, then the probability of failure can be computed by substituting the expression of Eq. (4.16) for the reliability index in Eq. (4.10). This presumes that the limit-state function is linear in the X-space and that **X** is a Gaussian vector. In any other case, Eq. (4.10) becomes an approximation of the probability of failure.

In the general case, where $g(\mathbf{X})$ is not linear in **X** and **X** is an arbitrary non-Gaussian vector, the reliability index of Eq. (4.16) can be computed by the Euclidean norm of the so-called design point $\mathbf{u}^*$, defined as the solution of the following equality-constrained quadratic optimization problem [20] [note the equivalence with Eq. (4.16)]:

$$\mathbf{u}^* = \arg\min\left\{\frac{1}{2}\mathbf{u}^T\mathbf{u} \,\Big|\, G(\mathbf{u}) = 0\right\} \qquad (4.17)$$

The design point $\mathbf{u}^*$ is located on the limit-state surface $G(\mathbf{U}) = 0$ and has minimum distance from the origin of the standard normal space, i.e. the mean value of **U**. Due to the rotational symmetry of the n-dimensional standard normal PDF $\varphi_n(.)$, the point $\mathbf{u}^*$ has the highest probability density of any realizations in the failure domain at the U-space. Therefore, $\mathbf{u}^*$ is also referred to as the most probable failure point. It is then obvious that the area in the vicinity of $\mathbf{u}^*$ will have the largest contribution to the integral in Eq. (4.1), which makes $\mathbf{u}^*$ the optimal point for the linearization of the limit-state function in the U-space and Eq. (4.10) the optimal first order approximation of the probability of failure, in the absence of specific knowledge on the shape of the surface $G(\mathbf{U}) = 0$.

The linearization of $G(\mathbf{U})$ at $\mathbf{u}^*$ is written as:

$$G(\mathbf{u}) \cong G_1(\mathbf{u}) = \nabla G(\mathbf{u}^*)^T (\mathbf{u} - \mathbf{u}^*) \tag{4.18}$$

where $\nabla G(\mathbf{u})$ is the gradient of of $G(\mathbf{u})$ with respect to $\mathbf{u}$. Let us denote by $\boldsymbol{\alpha}$, the unit normal vector to the hyperplane described by $G_1(\mathbf{u}) = 0$, pointing to the origin of the U-space, i.e.:

$$\boldsymbol{\alpha} = \frac{\nabla G(\mathbf{u}^*)}{\|\nabla G(\mathbf{u}^*)\|} \tag{4.19}$$

The significance of the vector $\boldsymbol{\alpha}$ will be discussed in Section 4.3.3. The design point can then be expressed as:

$$\mathbf{u}^* = -\beta \boldsymbol{\alpha} \tag{4.20}$$

Using Eqs. (4.19) and (4.20), we can rewrite Eq. (4.18) as:

$$G_1(\mathbf{u}) = \|\nabla G(\mathbf{u}^*)\|(\beta + \boldsymbol{\alpha}^T \mathbf{u}) \tag{4.21}$$

From Eq. (4.21), we obtain the following expression for the reliability index:

$$\beta = \frac{\mu_{G_1}}{\sigma_{G_1}} \tag{4.22}$$

whereby μ_{G_1}, σ_{G_1} are the mean and standard deviation of the linearization $G_1(\mathbf{u})$ of the limit-state function at the design point. Eq. (4.22) implies that the reliability index computed through Eq. (4.17) coincides with the reliability index of Eq. (4.11) for the linearized problem in the U-space.

The first order approximation of the probability of failure by the first order reliability method (FORM) is depicted in Figure 4.2. The following section discusses a number of optimization algorithms that have been implemented for the solution of the program of Eq. (4.17).

4.3 The first order reliability method 91

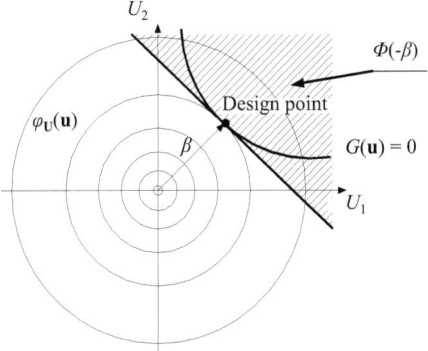

Figure 4.2: Reliability index and FORM approximation at the standard normal space.

4.3.1 Optimization algorithms

The quadratic program of Eq. (4.17) is solved in this study by application of gradient-based optimization techniques, i.e. the methods described in this section are iterative procedures which require the evaluation of the limit-state function and its gradient at each step. The gradient evaluation can be performed within the finite element (FE) program, by application of the direct differentiation method ([26], [142]). However, this method requires alterations at the FE-code level, which is outside the scope of this study. Alternatively, the gradient can be evaluated numerically, e.g. by the finite difference method. In this study, the forward finite difference scheme has been used, which requires $n + 1$ numerical evaluations of the limit-state function per iteration step. As the limit-state function is defined in the original basic random variable space $\mathbf{X}$ with respect to the FE solution, the transformation operator $\mathbf{T}(.)$ and its Jacobian are used to compute the limit-state function and transform its gradient to the $\mathbf{U}$-space.

The sequential quadratic programming method

Let us denote the objective function of the quadratic program defined in Eq. (4.17) by $F(\mathbf{u})$, i.e. $F(\mathbf{u}) = 1/2\mathbf{u}^T\mathbf{u}$. The solution $\mathbf{u}^*$ satisfies the necessary Kuhn-Tucker conditions [78], associated with the Lagrangian $L(\mathbf{u}, \lambda) = F(\mathbf{u}) + \lambda G(\mathbf{u})$:

$$\begin{aligned}\nabla_u L\left(\mathbf{u}^*, \lambda^*\right) &= \mathbf{u}^* + \lambda^* \nabla G\left(\mathbf{u}^*\right) = 0 \\ G\left(\mathbf{u}^*\right) &= 0\end{aligned} \qquad (4.23)$$

where λ is the Lagrange multiplier and $\nabla_u L(\mathbf{u}, \lambda)$ denotes the gradient of $L(\mathbf{u}, \lambda)$ with respect to $\mathbf{u}$. Linearization of Eq. (4.23) at $(\mathbf{u}_k, \lambda_k)$ yields:

$$\begin{bmatrix} \nabla_u^2 L(\mathbf{u}_k, \lambda_k) & \nabla G(\mathbf{u}_k) \\ \nabla G(\mathbf{u}_k)^T & 0 \end{bmatrix} \begin{bmatrix} \mathbf{d}_k \\ \kappa_k \end{bmatrix} = -\begin{bmatrix} \mathbf{u}_k \\ G(\mathbf{u}_k) \end{bmatrix} \qquad (4.24)$$

where $\mathbf{u}_{k+1} = \mathbf{u}_k + \xi_k \mathbf{d}_k$, $\lambda_{k+1} = \lambda_k + \xi_k(\kappa_k - \lambda_k)$, $\mathbf{d}_k$ and κ_k are the search directions for $\mathbf{u}$ and λ respectively and ξ_k is the step-length. The solution of Eq. (4.24) is equivalent to solving the quadratic sub-program with objective function $1/2\mathbf{d}_k^T \nabla_u^2 L(\mathbf{u}_k, \lambda_k)\mathbf{d}_k + \mathbf{u}_k^T\mathbf{d}_k$ and equality constraint $G(\mathbf{u}_k) + \nabla G(\mathbf{u}_k)^T \mathbf{d}_k = 0$ [9]. This approach corresponds to the sequential quadratic programming (SQP) method [9].

In the current implementation of the method, the Hessian of the Lagrangian $\nabla_u^2 L(\mathbf{u}_k, \lambda_k)$ is successively approximated using first order information, applying the Broyden-Fletcher-Goldfarb-Shanno (BFGS) scheme [9]. The step-length is computed by performing a line search at each step for the minimization of the following augmented Lagrangian function ([74], [117]):

$$L_a(\mathbf{u}_k, \lambda_k) = F(\mathbf{u}_k) + \lambda_k G(\mathbf{u}_k) + \frac{1}{2}c_k G(\mathbf{u}_k)^2 \qquad (4.25)$$

The choice of c_k in Eq. (4.25) should be sufficiently large, such that the function $L_a(\mathbf{u}_k, \lambda_k)$ is convex near the value of $\mathbf{u}_k$ at each step k [117]. The line search is performed in a relaxed fashion, by applying the Armijo test [78] and thus demanding a sufficient reduction of $L_a(\mathbf{u}_k, \lambda_k)$. The Armijo

4.3 The first order reliability method

method computes the step-length $\xi_k = b^i$, where $b \in (0, 1)$, by searching for the minimum $i \in \mathbb{N}$, so that the following equation is satisfied:

$$L_a\left(\mathbf{u}_k + b^i \mathbf{d}_k, \lambda_k + b^i\left(\kappa_k - \lambda_k\right)\right) \leq L_a\left(\mathbf{u}_k, \lambda_k\right)$$
$$+ \alpha b^i \nabla L_a\left(\mathbf{u}_k, \lambda_k\right)^{\mathrm{T}} \begin{bmatrix} \mathbf{d}_k \\ \kappa_k - \lambda_k \end{bmatrix} \quad (4.26)$$

wherein α is an algorithmic parameter which must satisfy $\alpha \in (0, 1)$. Eq. (4.26) is a sufficient decrease condition for $L_a\left(\mathbf{u}_k, \lambda_k\right)$, provided that $[\mathbf{d}_k, \kappa_k - \lambda_k]^{\mathrm{T}}$ is a descent direction [78].

The HL-RF method

The HL-RF method was originally proposed by Hasofer and Lind [53] and later combined with the Rosenblatt transformation by Rackwitz and Fiessler [107] to account for distribution information. According to the HL-RF method, the design point is recursively approximated by the following expression:

$$\mathbf{u}_{k+1} = \frac{1}{\left|\nabla G(\mathbf{u}_k)\right|^2}\left[\nabla G(\mathbf{u}_k)^{\mathrm{T}} \mathbf{u}_k - G(\mathbf{u}_k)\right]\nabla G(\mathbf{u}_k) \quad (4.27)$$

The expression in Eq. (4.27) can be derived by requiring that the linearization of $G(\mathbf{u})$ at $\mathbf{u}_k$ be perpendicular to $\nabla G(\mathbf{u}_k)$ at $\mathbf{u}_{k+1}$. Also, it is trivial to show that the HL-RF method coincides with the SQP method in the case where the Hessian of the Lagrangian is approximated by the identity matrix and the step-length is chosen as $\xi_k = 1$.

The HL-RF method is widely used due to its simplicity and rapid convergence. Nevertheless, under certain conditions, this method may fail to converge [22]. Therefore, certain modifications of the method have been suggested aiming at the improvement of its robustness. An improved version of the method, named improved HL-RF (iHL-RF) method, has been proposed by Zhang and Der Kiureghian [143]. According to this approach, an optimal step-length is chosen, i.e. $\xi_k \neq 1$, and the design point is updated using the direction:

94   4 Finite element reliability assessment

$$\mathbf{d}_k = \frac{1}{|\nabla G(\mathbf{u}_k)|^2}\left[\nabla G(\mathbf{u}_k)^T \mathbf{u}_k - G(\mathbf{u}_k)\right]\nabla G(\mathbf{u}_k) - \mathbf{u}_k \qquad (4.28)$$

At each step, the direction computed by Eq. (4.28) is used to perform a line search, in order to determine the value of ξ_k which minimizes a merit function $m(\mathbf{u})$. The line search is performed in this study applying the Armijo rule, which in this case reads for $\xi_k = b^i$, $b \in (0, 1)$:

$$m\left(\mathbf{u}_k + b^i \mathbf{d}_k\right) \le m(\mathbf{u}_k) + ab^i \nabla m(\mathbf{u}_k)^T \mathbf{d}_k \qquad (4.29)$$

The merit function reads [143]:

$$m(\mathbf{u}_k) = \frac{1}{2}\mathbf{u}_k^T \mathbf{u}_k + c_k |G(\mathbf{u}_k)| \qquad (4.30)$$

The method is proved to be unconditionally convergent for a choice of $c_k > \|\mathbf{u}_k\|/\|\nabla G(\mathbf{u}_k)\|$ [143].

The gradient projection method

The gradient projection (GP) method [110] is based on a modification of the steepest descent method to account for constraints. According to this method, the search direction is taken at each step k as the projection of the negative gradient of the objective function on the tangent plane of the feasible set, i.e. in the current case the limit-state surface $G(\mathbf{U}) = 0$. Let $\mathbf{s}_k$ be the parallel and $\gamma \cdot \nabla G(\mathbf{u}_k)$ the perpendicular vector component to the constraint surface of the gradient of the objective function $F(\mathbf{u}) = \frac{1}{2}\mathbf{u}^T\mathbf{u}$, where $\gamma \in \mathbb{R}$. The gradient vector $\nabla F(\mathbf{u}_k) = \mathbf{u}_k$ may then be decomposed as:

$$\mathbf{u}_k = \gamma \cdot \nabla G(\mathbf{u}_k) + \mathbf{s}_k \qquad (4.31)$$

Multiplying both sides of Eq. (4.31) by $\nabla G(\mathbf{u}_k)^T$ and rearranging, we get:

$$\gamma = \frac{\nabla G(\mathbf{u}_k)^T \mathbf{u}_k}{\nabla G(\mathbf{u}_k)^T \nabla G(\mathbf{u}_k)} \qquad (4.32)$$

4.3 The first order reliability method

where we have used that $\nabla G(\mathbf{u}_k)^T \mathbf{s}_k = 0$. Substituting Eq. (4.32) into Eq. (4.31), we get:

$$\mathbf{s}_k = \left[\mathbf{I} - \frac{1}{\nabla G(\mathbf{u}_k)^T \nabla G(\mathbf{u}_k)} \nabla G(\mathbf{u}_k) \nabla G(\mathbf{u}_k)^T\right] \cdot \mathbf{u}_k = \mathbf{P} \cdot \mathbf{u}_k \quad (4.33)$$

where **I** is the identity matrix and **P** is called the projection matrix. Furthermore, the search direction, i.e. the projection of the negative gradient, is given by:

$$\mathbf{d}_k = -\mathbf{P} \cdot \mathbf{u}_k \quad (4.34)$$

Starting from a vector $\mathbf{u}_k$, which satisfies the constraint, we compute $\mathbf{u}_{k+1}^0$, moving on the direction $\mathbf{d}_k$ by a step-length ξ_k. To include the case where the limit-state surface is nonlinear in the U-space, the new point, which lies on the tangent plane of the constraint surface, is pulled back onto the limit-state surface using a Newton-like method. A graphical illustration of the GP method is shown in Figure 4.3.

The GP method presents similar convergence behavior to the HL-RF method in the case where a constant step-length ξ_k is chosen [22]. To circumvent this problem, an improved version of the GP method (iGP) has been proposed by the author [95]. This approach is based on the idea that the search direction should be perpendicular to the gradient of the objective function at the design point. To avoid oscillations around the design point, an adaptive step-length is chosen according to a function in terms of the angle θ between the search direction and the gradient vector $\nabla F(\mathbf{u}_k) = \mathbf{u}_k$ (see Figure 4.3):

$$f_s(\theta) = 1 - (1 - \xi_{\min}) e^{-\left(\frac{\theta - \pi/2}{\rho}\right)} \quad (4.35)$$

where $\xi_{\min}$ is the minimum value of the step-length. This function has the advantage that it is non-constant only in a small area in the vicinity of $\theta = \pi/2$. The size of this area depends on the choice of ρ. In this study the values of $\xi_{\min} = 0.1$ and $\rho = 0.1$ are used. These values have been tested against a number of examples and are shown to contribute to a robust and efficient convergence behavior of the GP algorithm. In Figure 4.4, a graphical representation of the step-length reduction function is shown.

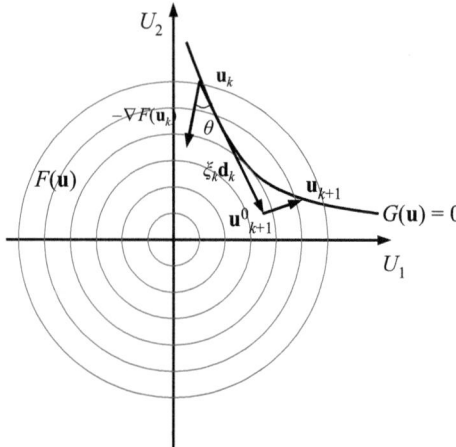

Figure 4.3: Graphical illustration of the gradient projection method.

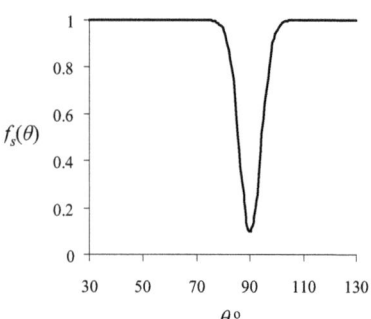

Figure 4.4: Graphical representation of the step-length reduction function.

4.3.2 Comparison of the optimization algorithms

This section compares the performance of the optimization algorithms presented in Section 4.3.1 for the solution of the FORM optimization problem for two examples.

Example 1

In the first example, the effect of numerical noise on the reliability assessment of a linear limit-state function, expressed in terms of two lognormal random variables, is examined. The noise is introduced by a trigonometric function, in which the amplitude and frequency are adjusted so that the failure probability is not influenced. The corresponding limit-state function is as follows

$$g(\mathbf{X}) = 1.0 - X_1/2000 - X_2/5000 + 0.001[\sin 0.1X_1 + \sin 0.1X_2] \quad (4.36)$$

The random variables in Eq. (4.36) have lognormal distributions with mean values $\mu_1 = 500$, $\mu_2 = 2000$ and standard deviations $\sigma_1 = 100$, $\sigma_2 = 400$ and they are statistically independent.

Starting from the mean point, all FORM algorithms were able to converge to a design point, apart from the HL-RF method. Table 4.1 shows the reliability index β computed and the required number of evaluations of the limit-state function, for the methods discussed in Section 4.3.1. The index i stands for the improved versions of the methods, i.e. the use of the merit function check for the HL-RF and the adaptive adjustment of the step-length utilizing the reduction function of Eq. (4.35) for the GP method. The latter proved to be the most efficient method for the present example, as it was able to converge with only 24 limit-state function evaluations.

Example 2

The second example, taken from [115], is chosen to test the performance of the FORM optimization algorithms for a nonlinear limit-state function, defined as:

$$g(\mathbf{X}) = X_1^3 + X_2^3 - 18.0 \tag{4.37}$$

where X_1, X_2 are statistically independent normal random variables. Two cases are considered for the moments of the random variables; (a) $\mu_1 = \mu_2 = 10$ and $\sigma_1 = \sigma_2 = 5$; (b) $\mu_1 = 10$, $\mu_2 = 9.9$ and $\sigma_1 = \sigma_2 = 5$.

In case (a) all algorithms were able to converge with similar performance in terms of efficiency (see Table 4.1). In case (b) the standard HL-RF and GP methods both exhibited unstable behaviors and failed to converge to a solution. On the other hand, their improved versions as well as the SQP method converged to the same reliability index, with the iGP proving to be more efficient than the iHL-RF and the SQP methods, as shown in Table 4.1.

Table 4.1: Comparison of the FORM algorithms.

Example	Method	β	Number of function evaluations
1	SQP	3.10108	36
	HL-RF	not converged	-
	iHL-RF	3.08807	64
	GP	3.08785	39
	iGP	3.08809	24
2(a)	SQP	2.24009	24
	HL-RF	2.24009	24
	iHL-RF	2.24009	24
	GP	2.24009	30
	iGP	2.24009	30
2(b)	SQP	2.22603	87
	HL-RF	not converged	-
	iHL-RF	2.22597	78
	GP	not converged	-
	iGP	2.22598	30

Based on the examples presented, we can conclude that the SQP, iHL-RF and iGP methods presented a robust convergence behavior. We should point out that the efficiency of the iGP method depends on the choice of ρ in Eq. (4.35), although the choice of $\rho = 0.1$ used in this study has been verified for a large number of examples cases. On the other hand, the iHL-RF and SQP methods, even though shown to be less efficient than the iGP for the examples considered here, are equipped with mathematical proofs of their convergence. A further discussion on the performance of FORM optimization algorithms can be found in [71] and [74]. Therein similar conclusions on the robustness of the iHL-RF and SQP methods are drawn.

It should be noted that the performance of the optimization methods considered in this study, as well as any gradient-based optimization method, depends on the quality of the approximation of the derivatives. As mentioned earlier, the derivatives of the limit-state function are evaluated here by application of the forward finite difference method. The performance of this method depends on the choice of the size of the finite difference step. For smooth functions, the error introduced by the finite difference approximation decreases as the step size becomes smaller. However, for nonsmooth functions a very small step size may introduce large errors in the derivatives. Implicit limit-state functions that depend on the outcome of a nonlinear FE calculation are usually nonsmooth (they include numerical noise). This is due to fact that the nonlinear FE solution is obtained by an iterative algorithm controlled by an error tolerance. In such cases, a relaxation of the tolerance of the FE algorithm may have a significant effect on the derivative approximation.

4.3.3 Sensitivity measures

Consider the linearization $G_1(\mathbf{u})$ of the limit-state function at the design point $\mathbf{u}^*$, given by Eq. (4.21). The variance of $G_1(\mathbf{u})$ can be expressed as follows [20]:

$$\sigma_{G_1}^2 = \left\|\nabla G(\mathbf{u}^*)\right\|^2 (\alpha_1^2 + \alpha_2^2 + \ldots + \alpha_n^2) = \left\|\nabla G(\mathbf{u}^*)\right\|^2 \qquad (4.38)$$

where $[\alpha_1, \alpha_2, \ldots, \alpha_n]$ are the elements of the vector $\boldsymbol{\alpha}$, defined in Eq. (4.19) – note that $\boldsymbol{\alpha}$ is of unit length. Eq. (4.21) implies that the squares of the elements of $\boldsymbol{\alpha}$ indicate the relative contribution of the corresponding elements of $\mathbf{U}$ to the total variance of the linearized limit-state function.

Therefore, the elements of $\boldsymbol{\alpha}$ provide measures of the relative influence of the equivalent standard normal variables $\{U_i, i = 1, ..., n\}$ ([8], [59]). The vector $\boldsymbol{\alpha}$ is often referred to as the vector of influence coefficients [35]. If the basic random variables $\mathbf{X}$ are statistically independent, then there is a one-to-one correspondence between $\mathbf{U}$ and $\mathbf{X}$ and the influence coefficients α_i also apply to $\mathbf{X}$. In this case and based on the definition in Eq. (4.19), a positive (resp. negative) value of α_i indicates that X_i is of capacity (resp. demand) type.

In the case where the components of $\mathbf{X}$ are statistically dependent, a one-to-one correspondence between $\mathbf{U}$ and $\mathbf{X}$ cannot be defined. Therefore, in this case the vector $\boldsymbol{\alpha}$ does not provide information about the relative importance of the basic variables $\mathbf{X}$. This problem can be resolved by expressing $\mathbf{U}$ as a linear function of an equivalent jointly Gaussian vector $\hat{\mathbf{X}}$ ([20], [55]):

$$\mathbf{u} = \mathbf{u}^* + \mathbf{J}_{\mathbf{u}^*,\mathbf{x}^*} \left(\hat{\mathbf{x}} - \mathbf{x}^* \right) \tag{4.39}$$

where $\mathbf{x}^*$ is the design point at the basic random variable space. Eq. (4.39) is derived by linearization of the transformation $\mathbf{U} = \mathbf{T}(\mathbf{X})$ at the design point and $\mathbf{J}_{\mathbf{u}^*,\mathbf{x}^*}$ is the Jacobian of the transformation at the design point [55]. The covariance matrix of $\hat{\mathbf{X}}$ reads:

$$\boldsymbol{\Sigma}_{\hat{\mathbf{x}}\hat{\mathbf{x}}} = \mathbf{J}_{\mathbf{u}^*,\mathbf{x}^*}^{-1} \mathbf{J}_{\mathbf{u}^*,\mathbf{x}^*}^{-T} \tag{4.40}$$

Substitution of Eq. (4.39) to Eq. (4.21) leads to:

$$G_1(\mathbf{u}) = \|\nabla G(\mathbf{u}^*)\| \boldsymbol{\alpha}^T \mathbf{J}_{\mathbf{u}^*,\mathbf{x}^*} \left(\hat{\mathbf{x}} - \mathbf{x}^* \right) \tag{4.41}$$

The variance of $G_1(\mathbf{u})$ is now written as:

$$\sigma_{G_1}^2 = \|\nabla G(\mathbf{u}^*)\|^2 \boldsymbol{\alpha}^T \mathbf{J}_{\mathbf{u}^*,\mathbf{x}^*} \boldsymbol{\Sigma}_{\hat{\mathbf{x}}\hat{\mathbf{x}}} \mathbf{J}_{\mathbf{u}^*,\mathbf{x}^*}^T \boldsymbol{\alpha} \tag{4.42}$$

Based on Eq. (4.42), a measure of the influence of the variances of $\hat{\mathbf{X}}$ (i.e. the diagonal terms of $\boldsymbol{\Sigma}_{\hat{\mathbf{x}}\hat{\mathbf{x}}}$) on the variance of $G_1(\mathbf{u})$ is defined as follows [55]:

$$\gamma = \frac{\boldsymbol{\alpha}^{T} \mathbf{J}_{\mathbf{u}^{*},\mathbf{x}} \cdot \mathbf{D}_{\hat{\mathbf{x}}}}{\left\| \boldsymbol{\alpha}^{T} \mathbf{J}_{\mathbf{u}^{*},\mathbf{x}} \cdot \mathbf{D}_{\hat{\mathbf{x}}} \right\|} \tag{4.43}$$

wherein $\mathbf{D}_{\hat{\mathbf{x}}} = \text{diag}\left[\sigma_{\hat{X}_i}\right]$. The sensitivity measure of Eq. (4.43) can be considered an approximate relative measure of the importance of the components of **X**. A positive (resp. negative) value of γ_i indicates that X_i is of capacity (resp. demand) type. It is straightforward to show that if the components of **X** are statistically independent, Eq. (4.43) reduces to the vector of influence coefficients **α**.

A number of additional sensitivity measures have been proposed focused on different aspects of the reliability problem. The omission sensitivity factors express the effect on the reliability index β if a basic random variable is replaced by a deterministic number [80]. Also, sensitivity measures of β can be defined in terms of its gradient with respect to parameters of the distribution of the basic variables, e.g. the moments of the marginal distributions, as well as with respect to deterministic parameters entering the definition of the limit-state function ([8], [35], [59]).

4.4 The inverse first order reliability method

In the structural design procedure, an optimal solution is sought, which minimizes a cost function and satisfies the reliability requirements. In practice, the reliability constraints are usually included in the design by application of global or partial safety factors (see Section 3.1). A more rigorous approach explicitly requires the design constraints to satisfy a certain reliability level. This approach is termed reliability-based design (RBD) (e.g. see [69]). In the case of only one design parameter, an efficient approach to solve the RBD problem based on an extension of the FORM has been proposed by Der Kiureghian et al. [27]. This approach is termed inverse FORM.

Let θ denote the design parameter. The inverse FORM optimization problem can be defined by extending the classical FORM problem of Eq.

(4.17) to include the design parameter θ and the additional constraint that $\beta = \beta_t$, where β_t is a target reliability index. For this single parameter case of the RBD problem, there is no cost function, since a unique value for θ can be determined to achieve the target reliability. Note that with the parameter θ undetermined, the limit-state function and, therefore, the design point and the reliability index are functions of θ. The objective is to select θ such that the minimum distance to the limit-state surface equals β_t. Next, an algorithm for the solution of the inverse FORM problem, based on an extension of the iHL-RF method is discussed.

The inverse HL-RF method

The inverse FORM problem can be described by the following formulation:

$$\|\mathbf{u}\| - \beta_t = 0 \tag{4.44}$$

$$\mathbf{u} + \frac{\|\mathbf{u}\|}{\|\nabla_{\mathbf{u}} G(\mathbf{u},\theta)\|} \nabla_{\mathbf{u}} G(\mathbf{u},\theta) = 0 \tag{4.45}$$

$$G(\mathbf{u},\theta) = 0 \tag{4.46}$$

where $\nabla_{\mathbf{u}}$ denotes the gradient operator with respect to $\mathbf{u}$ and $\|\mathbf{u}\| / \|\nabla_{\mathbf{u}} G(\mathbf{u}, \theta)\|$ is the Lagrange multiplier of the original FORM problem obtained by the HL-RF method. Eq. (4.44) expresses the reliability requirement of the RBD problem, while Eqs. (4.45) and (4.46) are the optimality conditions of the program of Eq. (4.17) for a fixed θ. The solution approach follows an iterative procedure, whereby at each iteration step k the unknowns $\mathbf{u}_k$, θ_k are updated according to the following rule:

$$\left[\mathbf{u}_{k+1}, \theta_{k+1}\right]^T = \left[\mathbf{u}_k, \theta_k\right]^T + \xi_k \mathbf{d}_k \tag{4.47}$$

where $\mathbf{d}_k$ is the search direction and ξ_k the step size. The search direction is obtained by solving the set of Eqs. (4.44)-(4.46) with $G(\mathbf{u}, \theta)$ in Eq. (4.46) substituted by linearization $G_1(\mathbf{u}, \theta)$ at $[\mathbf{u}_k, \theta_k]$:

$$G_1(\mathbf{u},\theta) = G(\mathbf{u}_k,\theta_k) + \nabla_\mathbf{u} G(\mathbf{u}_k,\theta_k)^T (\mathbf{u}-\mathbf{u}_k) + \frac{\partial G(\mathbf{u}_k,\theta_k)}{\partial \theta}(\theta-\theta_k) \quad (4.48)$$

The solution leads to the following expression for $\mathbf{d}_k$:

$$\mathbf{d}_k = \begin{bmatrix} -\beta_t \dfrac{\nabla_\mathbf{u} G(\mathbf{u}_k,\theta_k)}{\|\nabla_\mathbf{u} G(\mathbf{u}_k,\theta_k)\|} - \mathbf{u}_k \\ \dfrac{\nabla_\mathbf{u} G(\mathbf{u}_k,\theta_k)^T \mathbf{u}_k - G(\mathbf{u}_k,\theta_k) + \beta_t \|\nabla_\mathbf{u} G(\mathbf{u}_k,\theta_k)\|}{\partial G(\mathbf{u}_k,\theta_k)/\partial \theta} \end{bmatrix} \quad (4.49)$$

The step size ξ_k is chosen in this study by performing a line search at each iteration step for the minimization of the following merit function [93]:

$$m(\mathbf{u},\theta) = m_1(\mathbf{u},\theta) + m_2(\mathbf{u},\theta) \quad (4.50)$$

where $m_1(\mathbf{u}, \theta)$ is the merit function used in the iHL-RF algorithm [see Eq. (4.30)] and $m_2(\mathbf{u}, \theta)$ is the merit function for satisfying Eq. (4.44) [27]:

$$m_2(\mathbf{u},\theta) = \frac{1}{2}(\|\mathbf{u}\| - \beta_t)^2 \quad (4.51)$$

The line search is performed applying the Armijo rule, following Eq. (4.29).

4.5 The second order reliability method

In the case where the limit-state surface $G(\mathbf{u}) = 0$ is strongly nonlinear, the FORM approximation of the probability of failure using the linearization $G_1(\mathbf{u})$ at the design point can be highly erroneous. A solution to this problem may be given by the second order reliability method (SORM), which involves a second-order approximation of the limit-state function at the design point [39]. Consider the second-order Taylor series expansion of $G(\mathbf{u})$ at the design point $\mathbf{u}^*$:

$$G_2(\mathbf{u}) = \nabla G(\mathbf{u}^*)^T (\mathbf{u} - \mathbf{u}^*) + \frac{1}{2}(\mathbf{u} - \mathbf{u}^*)^T \nabla^2 G(\mathbf{u}^*)(\mathbf{u} - \mathbf{u}^*)$$
$$= \|\nabla G(\mathbf{u}^*)\| \left[\beta + \boldsymbol{\alpha}^T \mathbf{u} + \frac{1}{2\|\nabla G(\mathbf{u}^*)\|} (\mathbf{u} - \mathbf{u}^*)^T \nabla^2 G(\mathbf{u}^*)(\mathbf{u} - \mathbf{u}^*) \right] \quad (4.52)$$

where β is the FORM reliability index, $\boldsymbol{\alpha}$ the vector of influence coefficients and $\nabla^2 G(\mathbf{u}^*)$ the Hessian matrix of $G(\mathbf{u})$ at the design point. Then a rotation of the coordinate system $\mathbf{v} = \mathbf{R}\mathbf{u}$ is carried out such that the v_n-th direction passes through the origin and the design point. Therefore, the design point in the V-space is $\mathbf{v}^* = [0 \ldots 0 \ \beta]^T$. Setting $G_v(\mathbf{v}) = G(\mathbf{R}^T\mathbf{v})/\|\nabla G(\mathbf{u}^*)\|$, we can express the second-order expansion of $G_v(\mathbf{v})$ as follows [20]:

$$G_{v2}(\mathbf{v}) = \beta - v_n + \frac{1}{2}(\mathbf{v} - \mathbf{v}^*)^T \mathbf{A}(\mathbf{v} - \mathbf{v}^*) \quad (4.53)$$

where $\mathbf{A} = \mathbf{R}\nabla^2 G(\mathbf{u}^*)\mathbf{R}^T / \|\nabla G(\mathbf{u}^*)\|$. Using the matrix $\mathbf{A}_{11}$ of dimensions $(n-1) \times (n-1)$, formed by the first $n-1$ rows and columns of $\mathbf{A}$, we can further approximate Eq. (4.53) as follows [20]:

$$G_{v2}(\mathbf{v}) \approx \beta - v_n + \frac{1}{2}\mathbf{v}_1^T \mathbf{A}_{11} \mathbf{v}_1 \quad (4.54)$$

where $\mathbf{v}_1$ is the vector containing the first $n-1$ elements of $\mathbf{v}$. A further rotation of $\mathbf{v}_1$ to the principal coordinates of $\mathbf{A}_{11}$ reduces Eq. (4.54) to:

$$G_{v2}(\mathbf{v}) \approx \beta - v_n + \frac{1}{2}\sum_{i=1}^{n-1} \kappa_i v_i'^2 \quad (4.55)$$

where κ_i are the eigenvalues of $\mathbf{A}_{11}$. Eq. (4.55) defines a parabola, tangent to the limit-state surface at the design point, with $\{\kappa_i, i = 1, \ldots, n-1\}$ being the principal curvatures of the limit-state surface at the design point.

The SORM approximation of the probability of failure is based on the approximation of the failure domain in the integral of Eq. (4.1) by the one defined by the parabolic expression of Eq. (4.55). The latter requires the evaluation of the Hessian $\nabla^2 G(\mathbf{u}^*)$ of the limit-state function at the design point. A numerical evaluation of the Hessian by application of the finite difference method requires $n(n + 1)/2$ additional calls to the FE solver. Alternatively, the BFGS method [9] can be used for the approximation of the Hessian using first-order information within the optimization procedure for the evaluation of the design point [106]. This approach gives a fairly good approximation of the Hessian for small dimensions. Another method for the approximation of the Hessian within the context of the FORM optimization procedure is given in [22].

An expression of the SORM probability of failure, asymptotically exact for large values of β, is given by [11]:

$$P_f \approx \Phi(-\beta)\prod_{i=1}^{n-1}(1+\beta\kappa_i)^{-1/2} \approx \Phi(-\beta)\prod_{i=1}^{n-1}\left[1+\frac{\varphi(\beta)}{\Phi(-\beta)}\kappa_i\right]^{-1/2} \qquad (4.56)$$

with the second formula being slightly better than the first [106]. Eq. (4.56) gives a good approximation for values of $\beta \geq 1$. Otherwise, Tvedt's formula [136]:

$$P_f = \frac{1}{2} - \frac{1}{\pi}\int_0^\infty \sin\left(\beta z + \frac{1}{2}\sum_{i=1}^{n-1}\tan^{-1}(-\kappa_i z)\right)\frac{\exp\left(-\frac{1}{2}z^2\right)}{z\left[\prod_{i=1}^{n-1}(1+\kappa_i^2 z^2)\right]^{1/4}}dz \qquad (4.57)$$

should be used, which gives an exact expression for the probability integral in the domain defined by Eq. (4.55) in terms of a single-fold integral.

4.6 System reliability using FORM

This section presents the adopted approach for the solution of system reliability problems, i.e. problems which involve multiple limit-state functions (see Section 3.4), using FORM. Consider a set of limit-state functions $\{g_k(\mathbf{X}), k = 1,\ldots,m\}$ and the corresponding limit-state functions in the equivalent standard normal space $\{G_k(\mathbf{U}), k = 1,\ldots,m\}$. For each function $G_k(\mathbf{U})$, a component reliability problem can be defined for which the FORM optimization problem can be solved for the corresponding design point $\mathbf{u}_k^*$. The linearization of each component limit-state function $G_k(\mathbf{U})$ at the corresponding design point $\mathbf{u}_k^*$ reads:

$$G_{k1}(\mathbf{u}) = \|\nabla G_k(\mathbf{u}_k^*)\|(\beta_k + \boldsymbol{\alpha}_k^T \mathbf{u}) \tag{4.58}$$

where $\boldsymbol{\alpha}_k = \nabla G(\mathbf{u}_k^*)/\|\nabla G(\mathbf{u}_k^*)\|$ and $\beta_k = -\boldsymbol{\alpha}_k^T \mathbf{u}_k^*$ is the FORM component reliability index. Using the linearizations of Eq. (4.58), we can approximate the failure domain of the system reliability problem by a hyper-polygon. Although this approximation is a good choice for series systems, for parallel systems a better choice involves linearization at the so-called joint design point [20]. However, the component design points β_k are often used in both cases, since they are much easier to obtain.

Consider the random vector $\mathbf{Y} = [Y_1, ..., Y_m]^T$, defined by $Y_k = \boldsymbol{\alpha}_k^T \mathbf{U}$, $k = 1, ..., m$. The vector $\mathbf{Y}$ is jointly Gaussian with zero means, unit standard deviations and correlation coefficients $\rho_{y_k y_l} = \boldsymbol{\alpha}_k^T \boldsymbol{\alpha}_l$, $k, l = 1, ..., m$. For a series system, the first-order approximation of the failure probability reads [58]:

$$\begin{aligned} P_{f,ser} &\approx P\left[\bigcup_{k=1}^m \{y_k \geq \beta_k\}\right] \\ &= 1 - P\left[\bigcap_{k=1}^m \{y_k < \beta_k\}\right] \\ &= 1 - \Phi_m(\mathbf{B}, \mathbf{R}_{YY}) \end{aligned} \tag{4.59}$$

where $\Phi_m(.)$ is the m-variate standard normal cumulative distribution function, $\mathbf{R}_{YY}$ is the correlation matrix of $\mathbf{Y}$ and $\mathbf{B} = [\beta_1, ..., \beta_m]^T$. For a parallel system, the probability approximation reads [58]:

$$\begin{aligned} P_{f,par} &\approx P\left[\bigcap_{k=1}^m \{y_k \geq \beta_k\}\right] \\ &= P\left[\bigcap_{k=1}^m \{y_k < -\beta_k\}\right] \\ &= \Phi_m(-\mathbf{B}, \mathbf{R}_{YY}) \end{aligned} \tag{4.60}$$

Eqs. (4.59) and (4.60) both require the evaluation of the multi-normal distribution function, which can be done by use of approximations ([43], [91], [133]) or simulation methods [1].

The general system reliability problem given in Eq. (3.21) can be approached as follows. We first denote the failure events corresponding to the K parallel system problems by $\{C_k, k = 1,\ldots,K\}$. Then, Eq. (3.21) takes the form:

$$P_{f,sys} = P\left(\bigcup_{k=1}^{K} C_k\right) \quad (4.61)$$

Applying Eq. (2.11) recursively to Eq. (4.61) we obtain the following expression, also know as the Poincaré formula:

$$P_{f,sys} = \sum_{k=1}^{K} P(C_k) - \sum_{l=2}^{K}\sum_{k=1}^{l-1} P(C_k \cap C_l) + \ldots + (-1)^{K} P(C_1 \cap \ldots \cap C_K) \quad (4.62)$$

Each term in Eq. (4.62) represents a parallel system reliability problem, which can be approximated by Eq. (4.60). However, the parallel system reliability problems in Eq. (4.62) increase rapidly as K increases. Alternatively to Eq. (4.62), the narrow system reliability bounds given in [34] can be used. These bounds are based on low-order probabilities, thus avoiding the solution of parallel systems with a large number of components.

It should be noted that the FORM approximation for series systems can also be applied to the case where a component reliability problem has multiple local or global minima ([21], [35]). In such cases, an approximation of the failure domain by multiple linearizations will result in an improved FORM approximation of the probability of failure.

4.7 Simulation methods

As noted in Section 3.3.1, simulation methods estimate the probability of failure based on random sampling of the basic random variables. These methods are asymptotically exact, since they do not imply any approximation. The quality of the resulting estimate of the probability of failure is usually controlled by a variance estimate. It should be noted that unlike the FORM, simulation methods rarely provide information concerning the sensitivity of the failure probability with respect to the basic random variables.

The simulation methods described in this section have been implemented in the equivalent standard normal space **U**, derived from the marginal isoprobabilistic transformation discussed in Section 4.2. Also, as discussed in Section 3.4, the general system reliability problem can be expressed in terms of one equivalent limit-state function [see Eq. (3.22)], which can be formulated in the **U**-space. Therefore, the simulation methods will be discussed for the solution of the reliability problem of Eq. (4.1), taking into account that any system or component reliability problem can be reduced to an identical form – note that such a formulation is not suited for application of the FORM (resp. SORM) for the solution of system reliability problems, due to the nature of the FORM (resp. SORM) approximation.

4.7.1 Generation of random samples

This section briefly comments on the generation of samples of random variables, which is an essential part for most of the methods described in the following. The classical algorithms for the generation of random numbers produce samples that are relatively independent and asymptotically uniformly distributed in [0, 1]. These algorithms are based on recursive functions and usually need to be initialized by a seed. In principle, the samples produced are deterministic periodic sequences which can be reproduced if we apply the same seed. Therefore, the generated samples are called pseudo-random numbers. For an overview on algorithms for generation of pseudo-random numbers, the reader is referred to [71].

Simulation methods require the generation of samples from a prescribed PDF. This can be done by application of the transformation method [112]. Let U be a random variable uniformly distributed in [0, 1]. Taking into account that the distribution function of U is the identity

4.7 Simulation methods

function, i.e. $F_U(u) = u$, it is straightforward to show that the random variable X, defined by:

$$X = F_X^{-1}(U) \tag{4.63}$$

has the distribution function $F_X(x)$, provided that $F_X(x)$ is continuous and strictly increasing. Therefore, to simulate X it is sufficient to generate a sample u_i of U and then apply $x_i = F_X^{-1}(u_i)$ (see Figure 4.5). In the case where there is no explicit expression for the inverse of the distribution function, the inversion can be performed by application of an iterative procedure; alternatively approximate expressions can be used if such exist. Another way to circumvent this problem is to use an auxiliary distribution to generate the samples and then test the sample to determine whether it is accepted or rejected. This approach is called rejection-acceptance method [139].

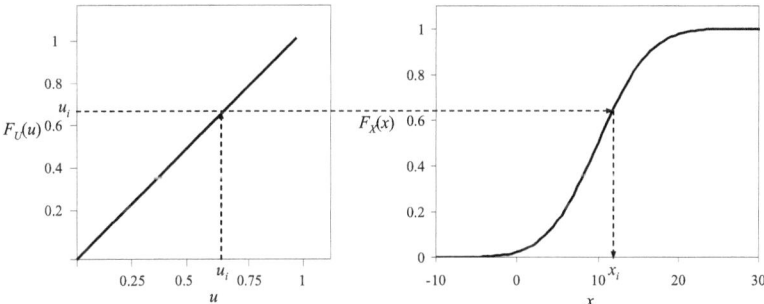

Figure 4.5: Generation of a sample from a Gaussian distribution by application of the transformation method.

Since the simulation methods are implemented in the standard normal space, the generated random samples should follow the independent standard normal distribution. Generation of samples of the standard Gaussian distribution can be done by application of the Box-Müller method [10]. Let U_1, U_2 be independent random variables uniformly distributed in [0, 1]. Let us then defined the following transformation:

$$X_1 = \sqrt{-2\ln U_1} \sin(2\pi U_2) \tag{4.64}$$

$$X_2 = \sqrt{-2\ln U_1} \cos(2\pi U_2) \tag{4.65}$$

If we apply Eq. (2.85) and taking into account that the joint PDF of $\mathbf{U} = [U_1, U_2]^T$ is $f_\mathbf{U}(\mathbf{u}) = 1$, it is easy to show that the vector $\mathbf{X} = [X_1, X_2]^T$ is an independent standard normal vector. Therefore, independent standard Gaussian variables can be simulated by drawing samples from independent uniform variables and applying the transformation of Eqs. (4.64)-(4.65).

4.7.2 The Monte Carlo method

The probability of failure in Eq. (4.1) can be rewritten in the following form:

$$P_f = \int_{G(\mathbf{U}) \leq 0} \varphi_n(\mathbf{u}) d\mathbf{u} = \int_{D_\mathbf{U}} I(\mathbf{u}) \varphi_n(\mathbf{u}) d\mathbf{u} = \mathrm{E}[I(\mathbf{u})] \tag{4.66}$$

where $D_\mathbf{U} = \mathbb{R}^n$, $\mathrm{E}[.]$ is the mathematical expectation operator and $I(\mathbf{u})$ is the indicator function, defined as:

$$I(\mathbf{u}) = \begin{cases} 1 & \text{if } G(\mathbf{u}) \leq 0 \\ 0 & \text{if } G(\mathbf{u}) > 0 \end{cases} \tag{4.67}$$

The probability of failure can then be estimated by generating N independent samples $\{\mathbf{u}_k \ (k = 1,\ldots,N)\}$ of $\varphi_n(\mathbf{u})$ and taking the sample mean of $I(\mathbf{u})$ [112], i.e.

$$\hat{P}_f = \hat{\mathrm{E}}[I(\mathbf{u})] = \frac{1}{N} \sum_{k=1}^{N} I(\mathbf{u}_k) \tag{4.68}$$

Eq. (4.68) gives an unbiased estimator for P_f, since $\mathrm{E}[\hat{P}_f] = P_f$. The variance of the estimator reads:

4.7 Simulation methods

$$\operatorname{Var}(\hat{P}_f) = \operatorname{Var}\left[\frac{1}{N}\sum_{k=1}^{N} I(\mathbf{u}_k)\right] = \frac{1}{N}\operatorname{Var}[I(\mathbf{u})]$$
$$= \frac{1}{N}\left(\operatorname{E}[I(\mathbf{u})^2] - \operatorname{E}[I(\mathbf{u})]^2\right) \quad (4.69)$$

where we have used that the random variables $I(\mathbf{u}_k)$ come from independent samples of identically distributed random variables. By noting that $P_f = \operatorname{E}[I(\mathbf{u})]$ and since $I(\mathbf{u})^2 = I(\mathbf{u})$ we can write Eq. (4.69) as:

$$\operatorname{Var}(\hat{P}_f) = \frac{1}{N} P_f (1 - P_f) \quad (4.70)$$

The coefficient of variation of the estimate is then expressed as:

$$CV_{\hat{P}_f} = \frac{\sqrt{\operatorname{Var}(\hat{P}_f)}}{P_f} = \sqrt{\frac{1 - P_f}{NP_f}} \approx \frac{1}{\sqrt{NP_f}} \quad (4.71)$$

where the approximation holds for small values of P_f. Therefore, if the magnitude of P_f is 10^{-k} and the target coefficient of variation is 0.1, then the required number of samples is $N = 10^{k+2}$. This means that for small failure probabilities, which is typically the case in civil engineering applications, the required number of samples is very high. Considering that each sample corresponds to a numerical evaluation of the limit-state function through a FE calculation, the Monte Carlo method is very inefficient for FE reliability analysis.

In practice, the coefficient of variation of $\hat{P}_f$ is estimated using the generated samples and serves as an error measure for the termination of the simulation. Usually a target coefficient of variation in the order of 0.05-0.1 is used. Consider the reliability analysis of a quadratic limit-state function in terms of a two-dimensional independent standard normal vector $\mathbf{U}$ (see also Figure 4.6):

$$G(\mathbf{U}) = 0.1(U_1^2 + U_2^2 - 2U_1 U_2) - \frac{1}{\sqrt{2}}(U_1 + U_2) + 2.5 \quad (4.72)$$

112   4 Finite element reliability assessment

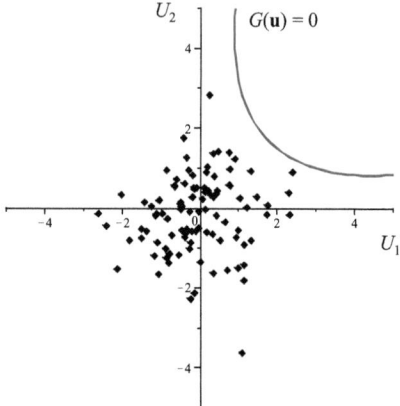

Figure 4.6: Graphical representation of the Monte Carlo method in a 2D standard normal space (100 samples).

As it can be seen in Figure 4.6, the probability of failure is relatively small – for 100 samples there is no hit in the failure domain. For a sample size of $N = 10^5$, the estimate of the probability of failure is $\hat{P}_f = 4.34 \cdot 10^{-3}$ with a coefficient of variation of 5%.

Next, two methods for the generation of samples, that serve as alternatives to pseudo-random sampling, are discussed.

Latin hypercube sampling

Latin hypercube sampling ([82], [90]) is a method for generating samples that are better distributed in the sample space compared to those obtained by pseudo-random number generators. According to this approach, the range of each random variable is divided into N sets with equal probability mass, where N is the total number of samples. Then, one sample is generated from each of the N sets. To simulate a random vector $\mathbf{u}_k$, one of the N sample values is randomly picked for each random variable. The advantage of this method is that it ensures a good spread of the generated samples and, therefore, usually converges faster than the crude Monte Carlo method. However, since the construction of the samples requires the knowledge of the total number of samples, the coefficient of variation cannot be used as a convergence criterion in this context.

Quasi-random sampling

Quasi-random numbers, also called low-discrepancy sequences, are deterministic sequences that aim at uniformly filling the unit hypercube $[0,1]^n$ [89]. The discrepancy of a sequence is a measure of its uniformity and it can be computed by comparing the actual samples in a given volume of a multidimensional space, say $[0,1]^n$, with the samples that should be there assuming a uniform distribution. The principle of the construction of quasi-random sequences is to find sequences with small discrepancy. The generated samples resemble samples of the n-dimensional uniform distribution, however, unlike pseudo-random samples, they are not statistically independent. Several types of sequences have been proposed, e.g. the Halton [51], Sobol [125] and Niederreiter [88] sequences. Figure 4.7 compares the samples of the uniform distribution in $[0,1]^2$ produced by pseudo-random sampling, latin hypercube sampling and the Niederreiter quasi-random sequence with $N = 1000$ samples. This figure shows that the Niederreiter sequence presents a better uniformity compared to the random sampling techniques. Quasi-random samples of an arbitrary distribution can be constructed in terms of samples of the uniform distributions by application of the methods discussed in Section 4.7.1.

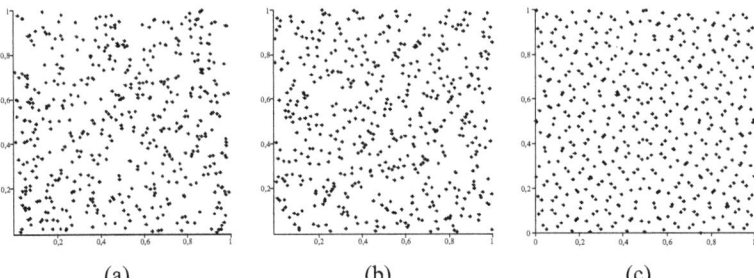

Figure 4.7: Sampling of a uniform distribution in $[0,1] \times [0,1]$. (a) Pseudo-random sampling. (b) Latin hypercube sampling. (c) Quasi-random sampling with Niderreiter sequence points.

4.7.3 Directional simulation

Directional simulation was first proposed by Deák [18] for the evaluation of multi-normal integrals and then extended for application to reliability analysis by Bjerager [7] and Ditlevsen et al. [36]. The n-dimensional independent standard normal random vector $\mathbf{U}$ can be expressed as $\mathbf{U} = R\mathbf{A}$, where R^2 is a χ^2-distributed random variable with n degrees of freedom and $\mathbf{A}$ is a random vector, independent of R and uniformly distributed on the n-dimensional unit hypersphere. The probability integral of Eq. (4.1) can then be expressed as follows:

$$P_f = \int_{G(\mathbf{U})\leq 0} \varphi_n(\mathbf{u}) d\mathbf{u} = \int_{D_\mathbf{A}} \int_{r(\mathbf{a})}^{\infty} f_R(s) ds \, f_\mathbf{A}(\mathbf{a}) d\mathbf{a} \qquad (4.73)$$

where $f_\mathbf{A}(\mathbf{a})$ is the PDF of $\mathbf{A}$, $f_R(s)$ is the PDF of R at a given direction $\mathbf{a}$, $D_\mathbf{A}$ is the unit hypersphere and $r(\mathbf{a})$ is the radius in the direction $\mathbf{a}$ to the boundary of the failure domain, i.e. $G[r(\mathbf{a})\mathbf{a}] = 0$, $r(\mathbf{a}) \in \mathbb{R}_+$ (see Figure 4.8). The nested integral in Eq. (4.73) can be evaluated analytically due to the knowledge of the distribution of R, i.e.

$$\int_{r(\mathbf{a})}^{\infty} f_R(s) ds = 1 - \int_0^{r(\mathbf{a})} f_R(s) ds = 1 - \chi_n^2\left(r(\mathbf{a})^2\right) \qquad (4.74)$$

where $\chi_n^2(.)$ is the CDF of the χ^2 distribution with n degrees of freedom. Therefore, Eq. (4.73) reduces to:

$$P_f = \int_{D_\mathbf{A}} \left[1 - \chi_n^2\left(r(\mathbf{a})^2\right)\right] f_\mathbf{A}(\mathbf{a}) d\mathbf{a} = \mathrm{E}\left[1 - \chi_n^2\left(r(\mathbf{a})^2\right)\right] \qquad (4.75)$$

An unbiased estimator of the probability of failure can then be obtained by generating N independent samples $\{\mathbf{a}_k \ (k = 1,\ldots,N)\}$ of the unit vector $\mathbf{A}$ and taking the sample mean of Eq. (4.75), i.e.

$$\hat{P}_f = \hat{\mathrm{E}}\left[1 - \chi_n^2\left(r(\mathbf{a})^2\right)\right] = \frac{1}{N}\sum_{k=1}^{N}\left[1 - \chi_n^2\left(r(\mathbf{a}_k)^2\right)\right] \qquad (4.76)$$

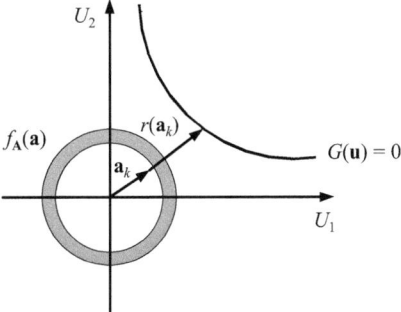

Figure 4.8: Graphical representation of the directional simulation in a 2D standard normal space.

A sample $\mathbf{a}_k$ of $\mathbf{A}$ can be simulated by generating an outcome $\mathbf{u}_k$ of $\mathbf{U}$ and setting $\mathbf{a}_k = \mathbf{u}_k/\|\mathbf{u}_k\|$. Alternatively, the samples can be generated by the rejection-acceptance method, whereby samples of the uniform distribution in $[-1, 1]^n$ are drawn and then rejected if outside of the unit hypersphere. Finally, the samples of $\mathbf{A}$ are computed by normalizing the accepted samples. The root of $G[r(\mathbf{a}_k)\mathbf{a}_k] = 0$ can be found by application of any line-search method. In this study, the secant method is applied. The variance of the estimator reads:

$$\operatorname{Var}\left(\hat{P}_f\right) = \operatorname{Var}\left\{\frac{1}{N}\sum_{k=1}^{N}\left[1 - \chi_n^2\left(r(\mathbf{a}_k)^2\right)\right]\right\}$$
$$= \frac{1}{N}\left(\operatorname{E}\left[\left(1 - \chi_n^2\left(r(\mathbf{a}_k)^2\right)\right)^2\right] - P_f^2\right)$$
(4.77)

The directional simulation method avoids the problem of the Monte Carlo method, namely that only a few samples fall in the failure domain. Moreover, due to the analytical evaluation of the probability associated with every sample, a single sample is sufficient for the evaluation of the exact probability corresponding to a hyperspherical failure surface in the U-space. This means that each sample value corresponds to a probability integral associated with a hyperspherical segment, the size of which depends on the total number of samples. Therefore, the directional simulation method is particularly efficient in the case of almost spherical failure surfaces. However, for nonspherical surfaces and a fixed sample size, the sample variance increases rapidly for increasing dimension n.

Directional simulation with deterministic directions

The directional simulation method assumes a uniform distribution of the random vector $\mathbf{A}$ on the unit hypersphere of the U-space. This means that each directional integral has an equivalent contribution to the probability of failure. Therefore, in the case where $\mathbf{A}$ is simulated randomly, a large number of samples are needed to provide a sufficiently good estimate of the probability. On the other hand, if a deterministic set of 'evenly' distributed points (directions) is used, the required sample size decreases considerably ([65], [86], [87], [95]).

The problem of distributing points equally on a hypersphere has attracted the attention of many scientists due to its application to a variety of scientific fields. Although it is not possible to find an exactly uniform distribution of points on a hypersphere, a number of methods have emerged which are able to provide sufficiently uniform distributions [113]. In this study, two methods have been implemented for this purpose.

The first method is based on equally dividing the spherical coordinates:

$$p = \{1, \omega_1, \omega_2, \ldots, \omega_{n-1}\} \qquad (4.78)$$

of the unit hypersphere, where $0 \leq \omega_i \leq \pi$ for $i \in \{1, \ldots, d-2\}$ and $0 \leq \omega_{n-1} \leq 2\pi$, and then transforming the derived vectors to Cartesian coordinates as follows ([65], [77]):

$$\begin{aligned}
a_1 &= \cos\omega_1 \\
a_2 &= \sin\omega_1 \cos\omega_2 \\
a_3 &= \sin\omega_1 \sin\omega_2 \cos\omega_3 \\
&\vdots \\
a_{n-1} &= \sin\omega_1 \sin\omega_2 \ldots \sin\omega_{n-2} \cos\omega_{n-1} \\
a_n &= \sin\omega_1 \sin\omega_2 \ldots \sin\omega_{n-2} \sin\omega_{n-1}
\end{aligned} \qquad (4.79)$$

The basic idea of this method is that the derived points define a tiling of the hypersphere by identical hypercubes of dimension $d - 1$ [77]. This geometric method, also referred to as hyperspace division method [86], has the advantage that the points are computed fast and their quality is parameter independent. For the generation of the points the algorithm described in [77] is used.

An alternative method, based on a physical approach, generates the points numerically by applying a potential minimizing principle in a set of particles with unit charges and forces of mutual repulsion [113]. The derived points are known as Fekete points ([86], [113]). These points are defined as the set $\{\mathbf{a}_1, \mathbf{a}_2, ..., \mathbf{a}_N\}^*$ that minimizes the Coulomb potential:

$$E = \sum_{1 \leq j \leq k \leq N} \left\| \mathbf{a}_j - \mathbf{a}_k \right\|^{-1} \tag{4.80}$$

In order to compute the points, an optimization algorithm [86] is applied starting from a random distribution of points on the hypersphere. According to this algorithm, every point is moved in each step on the direction defined by the tangent component of the total force acting on the point. The force vector is computed by taking the gradient of the Coulomb energy. For example, the total force acting on the point $\mathbf{a}_1$ is:

$$\frac{\partial}{\partial \mathbf{a}_1} \left(\sum_{1 \leq j \leq N} \left\| \mathbf{a}_1 - \mathbf{a}_j \right\|^{-1} \right) = \sum_{1 \leq j \leq N} \frac{\mathbf{a}_j - \mathbf{a}_1}{\left\| \mathbf{a}_1 - \mathbf{a}_j \right\|^3} \tag{4.81}$$

wherein the component derivatives are grouped in vector form. For details on the algorithm the reader is referred to [86]. The basic drawback of the method is that for high dimensions the generation of the Fekete points is time-consuming. In Figure 4.9, the distribution of 100 points on a unit sphere based on random sampling and the two described methods is shown. This figure shows that the geometric and physical method generate points much more uniformly distributed than the points based on random sampling.

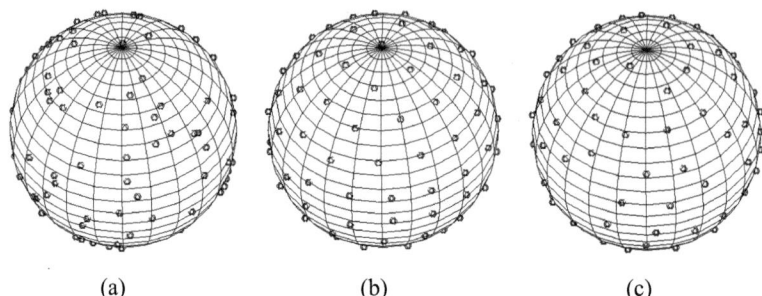

(a)   (b)   (c)

Figure 4.9: Distribution of 100 points on a sphere. (a) Random sampling. (b) Geometric points. (c) Fekete points.

4.7.4 Importance sampling methods

Importance sampling methods aim at producing samples (resp. directions) that are closer to the failure domain compared to the ones that occur according to the distribution of **U** (resp. **A**) [83]. The samples are produced based on an importance sampling function $h(\mathbf{u})$ [resp. $h(\mathbf{a})$]. The function $h(\mathbf{u})$ [resp. $h(\mathbf{a})$] is a PDF and is typically constructed based on previous information. A number of importance sampling methods are based on a given design point $\mathbf{u}^*$, i.e. the simulation is preceded by a FORM optimization. Note that the majority of these methods can also incorporate several design points and therefore can also be applied to series system reliability problems. Alternatively, an adaptive approach can be employed, according to which the sampling function is constructed based on results from a previous simulation. It is important to note that the support of the sampling function $h(\mathbf{u})$ [resp. $h(\mathbf{a})$] should include the support of the PDF of **U** (resp. **A**), or else the resulting estimate of the probability of failure might be biased.

Using the function $h(\mathbf{u})$ we can perform the integration in Eq. (4.1) by scaling the integrand, as follows [119]:

$$P_f = \int_{G(\mathbf{u}) \leq 0} \frac{\varphi_n(\mathbf{u})}{h(\mathbf{u})} h(\mathbf{u}) \, d\mathbf{u} = \int_{D_U} I(\mathbf{u}) \frac{\varphi_n(\mathbf{u})}{h(\mathbf{u})} h(\mathbf{u}) \, d\mathbf{u} = \mathrm{E}\left[I(\mathbf{u}) \frac{\varphi_n(\mathbf{u})}{h(\mathbf{u})} \right] \quad (4.82)$$

4.7 Simulation methods

where D_U is the support of $h(\mathbf{u})$ and $I(\mathbf{u})$ is the indicator function defined in Eq. (4.67). An unbiased estimator of P_f is given by:

$$\hat{P}_f = \hat{\mathrm{E}}\left[I(\mathbf{u})\frac{\varphi_n(\mathbf{u})}{h(\mathbf{u})}\right] = \frac{1}{N}\sum_{k=1}^{N} I(\mathbf{u}_k)\frac{\varphi_n(\mathbf{u}_k)}{h(\mathbf{u}_k)} \qquad (4.83)$$

wherein the samples $\{\mathbf{u}_k \; (k = 1,\ldots,N)\}$ are generated according to $h(\mathbf{u})$. The variance of the estimate is as follows:

$$\mathrm{Var}(\hat{P}_f) = \frac{1}{N}\left(\mathrm{E}\left[I(\mathbf{u})\left(\frac{\varphi_n(\mathbf{u})}{h(\mathbf{u})}\right)^2\right] - P_f^2\right) \qquad (4.84)$$

The optimal sampling density can be obtained by minimizing the variance of Eq. (4.84), which yields [119]:

$$h(\mathbf{u}) = \begin{cases} \dfrac{\varphi_n(\mathbf{u})}{P_f} & \text{if } G(\mathbf{u}) \leq 0 \\ 0 & \text{otherwise} \end{cases} \qquad (4.85)$$

This function is in general not practical, since it requires the knowledge of P_f. The function of Eq. (4.85) would require only one sample for the estimation of the exact value of the probability of failure.

Similar results can be obtained for the directional simulation method, using a directional importance sampling function $h(\mathbf{a})$ [7]. The integration in Eq. (4.75) can be performed as follows:

$$P_f = \int_{D_\mathbf{A}} \left[1 - \chi_n^2(r(\mathbf{a})^2)\right]\frac{f_\mathbf{A}(\mathbf{a})}{h(\mathbf{a})} h(\mathbf{a}) d\mathbf{a} = \mathrm{E}\left[\left(1 - \chi_n^2(r(\mathbf{a})^2)\right)\frac{f_\mathbf{A}(\mathbf{a})}{h(\mathbf{a})}\right] \qquad (4.86)$$

where $D_\mathbf{A}$ is the support of $h(\mathbf{a})$. An estimate of the probability of failure is then given as follows:

$$\hat{P}_f = \hat{\mathrm{E}}\left[\left(1 - \chi_n^2(r(\mathbf{a})^2)\right)\frac{f_\mathbf{A}(\mathbf{a})}{h(\mathbf{a})}\right] = \frac{1}{N}\sum_{k=1}^{N}\left[\left(1 - \chi_n^2(r(\mathbf{a}_k)^2)\right)\frac{f_\mathbf{A}(\mathbf{a}_k)}{h(\mathbf{a}_k)}\right] \qquad (4.87)$$

wherein the samples $\{\mathbf{a}_k \ (k = 1,\ldots,N)\}$ are generated according to $h(\mathbf{a})$. The variance of the estimate reads:

$$\mathrm{Var}\left(\hat{P}_f\right) = \frac{1}{N}\left(\mathrm{E}\left\{ \left[\left(1 - \chi_n^2\left(r(\mathbf{a})^2\right)\right)\frac{f_{\mathbf{A}}(\mathbf{a})}{h(\mathbf{a})}\right]^2 \right\} - P_f^2 \right) \quad (4.88)$$

Minimizing the variance of Eq. (4.88), we get the optimal directional sampling density [7]:

$$h(\mathbf{a}) = \frac{1 - \chi_n^2\left(r(\mathbf{a})^2\right)}{P_f} f_{\mathbf{A}}(\mathbf{a}) \quad (4.89)$$

The sampling density of Eq. (4.89) suffers the same problem as the one of Eq. (4.85), i.e. it requires the knowledge of P_f and therefore it is not of practical interest.

The following sections discuss a number of importance sampling methods which use different concepts for estimating an importance sampling density that yields a reduction of the variance of $\hat{P}_f$.

Standard importance sampling

In the standard importance sampling, the sampling function $h(\mathbf{u})$ is chosen as a standard normal PDF centered at the design point $\mathbf{u}^*$ [119], i.e.

$$h(\mathbf{u}) = \varphi_n(\mathbf{u} - \mathbf{u}^*) \quad (4.90)$$

Substituting Eq. (4.90) to Eq. (4.83), we get an unbiased estimator of P_f:

$$\hat{P}_f = \frac{1}{N}\sum_{k=1}^{N} I(\mathbf{u}_k) \frac{\varphi_n(\mathbf{u}_k)}{\varphi_n(\mathbf{u}_k - \mathbf{u}^*)} \quad (4.91)$$

wherein the samples $\{\mathbf{u}_k \ (k = 1,\ldots,N)\}$ are generated according to $\varphi_n(\mathbf{u} - \mathbf{u}^*)$. Standard importance sampling gives satisfactory results if the design point is well identified and there exist no additional minima. In the case of multiple design points or series system reliability analysis, the

4.7 Simulation methods

importance sampling function can be chosen as the following stratified sampling density [119]:

$$h_{sys}(\mathbf{u}) = \frac{\sum_{i=1}^{m} \Phi(-\beta_i) h_i(\mathbf{u})}{\sum_{i=1}^{m} \Phi(-\beta_i)} \qquad (4.92)$$

where $h_i(\mathbf{u})$ is the sampling density for each design point and β_i is the corresponding FORM reliability index.

In Figure 4.10, the standard importance sampling method is applied to the reliability analysis of the limit-state function of Eq. (4.72). In this figure, it is shown that for this sampling density almost half of the samples fall in the failure domain. For a sample size of $N = 10^3$, the estimate of the probability of failure is $\hat{P}_f = 4.38 \cdot 10^{-3}$ with a coefficient of variation of 5.8%.

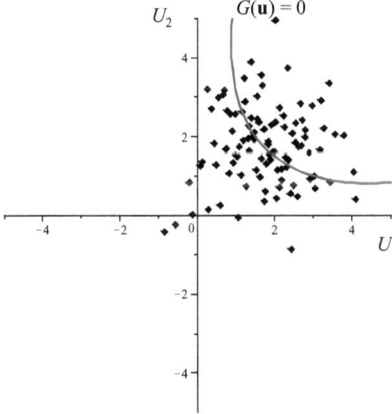

Figure 4.10: Graphical representation of the standard importance sampling method in a 2D standard normal space (100 samples).

Conditional sampling

The conditional sampling approach [52] is based on eliminating a large number of samples with zero probability of falling in the failure domain. Since the design point is the nearest point of the failure domain to the origin of the U-space, the hypersphere with center at the origin and radius equal to the reliability index β can be excluded from the sampling space. This can be done by choosing as sampling function the conditional standard normal PDF, given that $\|\mathbf{u}\| > \beta$, i.e.

$$h(\mathbf{u}) = \frac{1}{1 - \chi_n^2(\beta^2)} \varphi_n(\mathbf{u}) \qquad \mathbf{u} \in \{\mathbb{R}^n | \|\mathbf{u}\| > \beta\} \qquad (4.93)$$

where $\chi_n^2(.)$ is the CDF of the χ^2 distribution with n degrees of freedom. Substituting Eq. (4.93) to Eq. (4.83) we get an unbiased estimator of the failure probability:

$$\hat{P}_f = \left[1 - \chi_n^2(\beta^2)\right] \frac{1}{N} \sum_{k=1}^{N} I(\mathbf{u}_k) \qquad (4.94)$$

To simulate the vector $\mathbf{u}$ according to $h(\mathbf{u})$, we transform $\mathbf{U}$ to its polar coordinates $\mathbf{U} = R\mathbf{A}$, where $\mathbf{A}$ is a random vector uniformly distributed on the n-dimensional unit hypersphere and R^2 is distributed according to the truncated χ^2-distribution with n degrees of freedom, given that $R^2 > \beta^2$. The vector $\mathbf{A}$ can be simulated as discussed in Section 4.7.3. The variable R can be simulated by application of the rejection-acceptance procedure – an efficient approach is discussed in [52].

The conditional sampling method is particularly efficient in the case of hyperspherical failure surfaces. However, if the area in the vicinity of the design point has the largest contribution to the probability integral, then the conditional sampling method may produce a large number of unnecessary samples [see Figure 4.11(a)]. In such cases, it is suggested here that the conditional sampling be combined with the standard importance sampling method. According to this approach, we obtain the following sampling function:

4.7 Simulation methods

$$h(\mathbf{u}) = \frac{1}{1-C_I} \varphi_n(\mathbf{u}-\mathbf{u}^*) \quad \mathbf{u} \in \{\mathbb{R}^n | \|\mathbf{u}\| > \beta\} \tag{4.95}$$

The constant C_I is given by the following expression:

$$C_I = \int_{\|\mathbf{u}\| \leq \beta} \varphi_n(\mathbf{u}-\mathbf{u}^*)d\mathbf{u} \tag{4.96}$$

Unfortunately, the integral in Eq. (4.96) cannot be computed analytically. However, since the integration domain is bounded, we can compute the integral efficiently by application of any numerical approach. An efficient approach for problems of any dimension is to apply quasi-random sampling to the n-dimensional hypercube bounding the β-hypersphere. An unbiased estimate of the probability of failure is given by:

$$\hat{P}_f = [1-C_I]\frac{1}{N}\sum_{k=1}^{N} I(\mathbf{u}_k) \frac{\varphi_n(\mathbf{u}_k)}{\varphi_n(\mathbf{u}_k - \mathbf{u}^*)} \tag{4.97}$$

The samples $\mathbf{u}_k$ are generated by simulating $\varphi(\mathbf{u} - \mathbf{u}^*)$ and rejecting the samples for which $\|\mathbf{u}_k\| \leq \beta$.

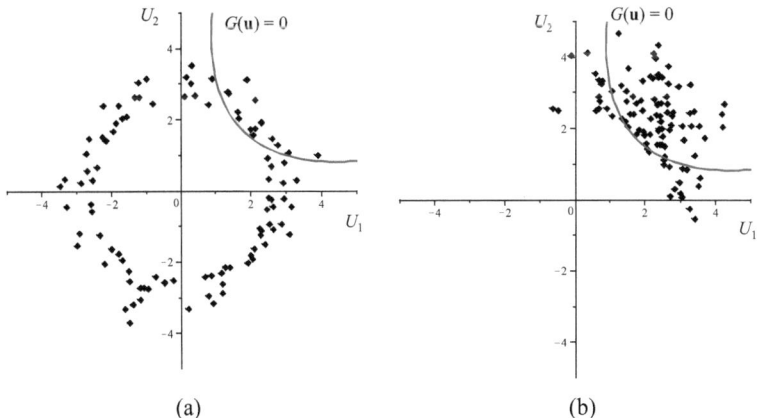

Figure 4.11: Graphical representation of conditional sampling methods in a 2D standard normal space (100 samples). (a) Conditional sampling. (b) Conditional importance sampling.

Figure 4.11 shows a graphical representation of the two conditional sampling approaches for the limit-state function of Eq. (4.72). It is shown that for this example the conditional importance sampling method performs better, since the majority of the samples fall in the failure domain. Also, comparing Figure 4.11(b) with Figure 4.10 we see that this approach outperforms the standard importance sampling method.

Axis orthogonal importance sampling

Another method based on the design point is the axis orthogonal importance sampling method [60]. In this method, the simulation takes place at the tangent hyperplane to the failure surface, centered at the design point. The sampling space is thus rotated and reduced by one dimension, i.e. $\mathbf{V} = \mathbf{R}\mathbf{U}$, $\mathbf{V} = [\mathbf{V}_1^T, V_n]^T$, where $\mathbf{V}_1$ contains the $(n-1)$-dimensional subspace, V_n defines the axis through the origin and the design point and $\mathbf{R}$ is a suitable rotation matrix. The probability integral can then be expressed as follows:

$$P_f = \int_{G(\mathbf{U}) \leq 0} \varphi_n(\mathbf{u}) \, d\mathbf{u} = \int_{D_{\mathbf{V}_1}} \int_{b(\mathbf{v}_1)}^{\infty} f_{V_n|\mathbf{V}_1}(v_n | \mathbf{v}_1) dv_n \, f_{\mathbf{V}_1}(\mathbf{v}_1) d\mathbf{v}_1 \quad (4.98)$$

where $f_{\mathbf{V}_1}(\mathbf{v}_1)$ is the $n-1$ standard normal PDF on the hyperplane $v_n = 0$, $f_{V_n|\mathbf{V}_1}(v_n|\mathbf{v}_1)$ is the conditional PDF of V_n given the vector $\mathbf{v}_1$, $D_{\mathbf{V}_1} = \mathbb{R}^{n-1}$ and $b(\mathbf{v}_1)$ is the distance to the failure surface in a direction orthogonal to the hyperplane at $\mathbf{v}_1$, i.e. the solution of $G[\mathbf{R}^T[\mathbf{v}_1^T, b(\mathbf{v}_1)]^T] = 0$ (see Figure 4.12). Due to the symmetry of $\varphi_n(\mathbf{u})$, the conditional PDF $f_{V_n|\mathbf{V}_1}(v_n|\mathbf{v}_1)$ is equal to the standard normal PDF centered at $\mathbf{v}_1$ and rotated to the V_n-axis. We thus have:

$$\int_{b(\mathbf{v}_1)}^{\infty} f_{V_n|\mathbf{V}_1}(v_n|\mathbf{v}_1) dv_n = \Phi(-b(\mathbf{v}_1)) \quad (4.99)$$

where $\Phi(.)$ is the standard normal CDF. Therefore, Eq. (4.98) reduces to:

$$P_f = \int_{D_{\mathbf{V}_1}} \Phi(-b(\mathbf{v}_1)) f_{\mathbf{V}_1}(\mathbf{v}_1) d\mathbf{v}_1 = \mathrm{E}\left[\Phi(-b(\mathbf{v}_1))\right] \quad (4.100)$$

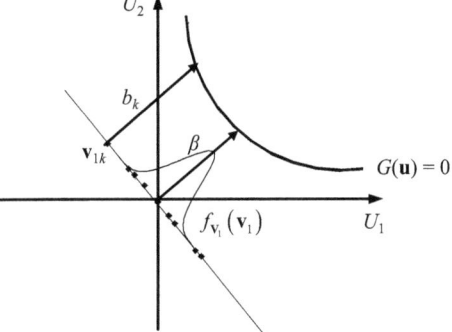

Figure 4.12: Graphical representation of the axis orthogonal importance sampling in a 2D standard normal space.

We can then compute an unbiased estimate of the failure probability by generating N samples $\{\mathbf{v}_{1k}\ (k=1,\ldots,N)\}$ of $\mathbf{V}_1$, solving $G[\mathbf{R}^T[\mathbf{v}_1^T, b_i]^T] = 0$ for b_i with any line-search method (e.g. the secant method) and applying:

$$\hat{P}_f = \frac{1}{N}\sum_{k=1}^{N}\Phi(-b_k) \qquad (4.101)$$

The axis orthogonal importance sampling method is shown to be very efficient in low-dimensional problems with failure surfaces in the vicinity of the design point. The method can be applied to high-dimensional problems with an initial crude Monte Carlo run to approximate the important direction V_n (e.g. see [68], [102] and [118]). However, the efficiency of the method decreases as the deviation of the direction V_n from the direction pointing to the design point increases.

Directional importance sampling

In the directional simulation method, the largest contribution to the failure probability comes from directions in the vicinity of the minimum $r(\mathbf{a})$, defined by the design point $\mathbf{u}^*$. The directional importance sampling method ([7], [36]) is based on concentrating the simulated directions around $\mathbf{u}^*$ by using as importance sampling function $h(\mathbf{a})$ the normalized standard normal distribution, truncated at the hyperplane tangent to the failure surface at $\mathbf{u}^*$. Assume that $\mathbf{v}$ is the direction pointing to the design point. The sampling function $h(\mathbf{a})$ can then be defined as follows [7]:

$$h(\mathbf{a}) = \frac{1-\chi_n^2\left[\left(\frac{\beta}{\mathbf{a}^T\mathbf{v}}\right)^2\right]}{\Phi[-\beta]} f_A(\mathbf{a}) \qquad \mathbf{a} \in \{D_A | \mathbf{a}^T\mathbf{v} > 0\} \qquad (4.102)$$

where $f_A(\mathbf{a})$ is the uniform PDF on the n-dimensional hypersphere D_A. We can then write Eq. (4.86) as follows:

$$P_f = \int_{\mathbf{a}^T\mathbf{v}>0} \left[1-\chi_n^2\left(r(\mathbf{a})^2\right)\right] \frac{f_A(\mathbf{a})}{h(\mathbf{a})} h(\mathbf{a}) \, d\mathbf{a}$$

$$= \Phi[-\beta] \mathrm{E}\left\{\frac{1-\chi_n^2\left(r(\mathbf{a})^2\right)}{1-\chi_n^2\left[\left(\frac{\beta}{\mathbf{a}^T\mathbf{v}}\right)^2\right]}\right\} \qquad (4.103)$$

An estimate of the probability of failure is then given as follows:

$$\hat{P}_f = \Phi[-\beta] \frac{1}{N} \sum_{k=1}^{N} \frac{1-\chi_n^2\left(r(\mathbf{a}_k)^2\right)}{1-\chi_n^2\left[\left(\frac{\beta}{\mathbf{a}_k^T\mathbf{v}}\right)^2\right]} \qquad (4.104)$$

A sample $\mathbf{a}_k$ of the function $h(\mathbf{a})$ can be generated using the following expression [36]:

$$\mathbf{a}_k = \frac{\mathbf{u}_k + \left(v_k - \mathbf{v}^T\mathbf{u}_k\right)\mathbf{v}}{\left\|\mathbf{u}_k + \left(v_k - \mathbf{v}^T\mathbf{u}_k\right)\mathbf{v}\right\|} \qquad (4.105)$$

where $\mathbf{u}_k$ is a sample of the standard normal vector $\mathbf{U}$ and v_k is a sample of the following truncated normal distribution:

$$f_V(v) = \frac{\varphi(v)}{\Phi(-\beta)} \qquad v \in (\beta, \infty) \qquad (4.106)$$

The sampling function of Eq. (4.102) is the optimal sampling function for any linear limit-state surface at the U-space [compare with Eq. (4.89)]. However, this function is zero for $\mathbf{a}^T\mathbf{v} \leq 0$ (see Figure 4.13). This means that the contribution of the directions pointing to this half-space is ignored and therefore the probability estimate of Eq. (4.104) may be biased. To circumvent this problem we can use a mixing sampling density [7], defined as follows:

$$h_m(\mathbf{a}) = \begin{cases} p \cdot f_A(\mathbf{a}) + (1-p) \cdot h(\mathbf{a}) & \text{for } \mathbf{a}^T\mathbf{v} > 0 \\ p \cdot f_A(\mathbf{a}) & \text{for } \mathbf{a}^T\mathbf{v} \leq 0 \end{cases} \quad (4.107)$$

where $p \in [0,1]$ is the mixing probability. The sampling function $h_m(\mathbf{a})$ produces samples according to the uniform density $f_A(\mathbf{a})$ with probability p and according to the sampling function $h(\mathbf{a})$ with probability $1 - p$.

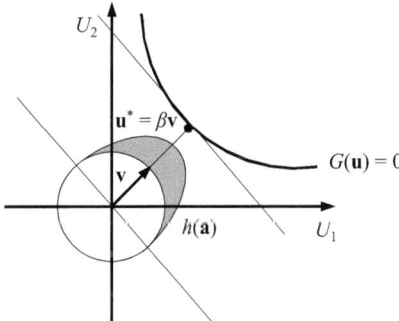

Figure 4.13: Graphical representation of the directional importance sampling in a 2D standard normal space.

Adaptive importance sampling

Several adaptive techniques have been proposed for the estimation of an optimal sampling density, based on a previous simulation (e.g. [3], [12]). These methods construct an importance sampling density $h_a(\mathbf{u})$ using the samples of the first simulation that fall in the failure domain. One approach is to perform the first simulation using an importance sampling function

$h(\mathbf{u})$ based on the design point, e.g. the function of Eq. (4.90). Let $\{\mathbf{u}_k \ (k = 1,\ldots,m)\}$ be the samples of the first simulation according to $h(\mathbf{u})$. The method described in [12] suggests using as sampling function the joint normal distribution, i.e.

$$h_a(\mathbf{u}) = \varphi_n\left(\mathbf{u} - \hat{\boldsymbol{\mu}}, \hat{\boldsymbol{\Sigma}}\right) \tag{4.108}$$

with mean value vector $\hat{\boldsymbol{\mu}}$ and covariance matrix $\hat{\boldsymbol{\Sigma}}$ determined as the mean and covariance of the samples from the first simulation, conditional on the failure domain:

$$\hat{\boldsymbol{\mu}} = \hat{\mathrm{E}}\left[\mathbf{u}|G(\mathbf{u}) \leq 0\right] = \frac{1}{m\hat{P}_{f1}} \sum_{k=1}^{m} \mathbf{u}_k I(\mathbf{u}_k) \frac{\varphi_n(\mathbf{u}_k)}{h(\mathbf{u}_k)} \tag{4.109}$$

$$\hat{\boldsymbol{\Sigma}} = \hat{\mathrm{E}}\left[(\mathbf{u} - \hat{\boldsymbol{\mu}})(\mathbf{u} - \hat{\boldsymbol{\mu}})^{\mathrm{T}}|G(\mathbf{u}) \leq 0\right]$$
$$= \frac{1}{m\hat{P}_{f1}} \sum_{k=1}^{m} (\mathbf{u}_k - \hat{\boldsymbol{\mu}})(\mathbf{u}_k - \hat{\boldsymbol{\mu}})^{\mathrm{T}} I(\mathbf{u}_k) \frac{\varphi_n(\mathbf{u}_k)}{h(\mathbf{u}_k)} \tag{4.110}$$

where $I(\mathbf{u})$ is the indicator function defined in Eq. (4.67) and $\hat{P}_{f1}$ is the estimate of the probability of failure from the first simulation. The sampling density $h_a(\mathbf{u})$ can be further adapted using results from subsequent simulations. Figure 4.14(a) shows samples generated using the sampling density of Eq. (4.108) for the limit-state function of Eq. (4.72) with the moments estimated from a simulation according to the standard importance sampling function of Eq. (4.90).

An alternative approach constructs the adaptive sampling density $h_a(\mathbf{u})$ by applying kernel density estimation using the samples of the first simulation that fall in the failure domain, say $\{\mathbf{u}_k \ (k = 1,\ldots,m_f)\}$ [3], i.e.

$$h_a(\mathbf{u}) = \frac{1}{m_f} \sum_{k=1}^{m_f} \frac{1}{w^n} K_n\left(\frac{\mathbf{u} - \mathbf{u}_k}{w}\right) \tag{4.111}$$

where $K_n(\mathbf{u})$ is the Gaussian kernel of dimension n, i.e. $K_n(\mathbf{u}) = \varphi_n(\mathbf{u}, \hat{\boldsymbol{\Sigma}})$, $\hat{\boldsymbol{\Sigma}}$ is the covariance of the samples $\{\mathbf{u}_k \ (k = 1,\ldots,m_f)\}$ and w controls the spread of each kernel. An optimal value of w can then be selected by

4.7 Simulation methods

minimizing the variance of the new estimator based on the samples of the first simulation [3]. Moreover, local spread parameters $\lambda_k w$ can be defined by computing $\lambda_k = [h_a(\mathbf{u}_k)]^{-\alpha}$, $\alpha \in [0,1]$, for $k = 1,\ldots,m_f$, and then constructing a new sampling density substituting w with $\lambda_k w$. Therefore, the kernels in areas with smaller density will have a wider spread, which will yield a smoother tail of $h_a(\mathbf{u})$. In Figure 4.14(b) samples generated by this method are shown. Comparing with Figure 4.14(a), we see that this method results in a wider spread of the generated samples.

In [5] it is proposed to combine the kernel density method with a generation of the initial failure points using Markov chain Monte Carlo simulation, starting from some given failure points – Markov chains are further discussed in Section 4.8.1. This approach is more suitable for application to domains with multiple design points or system reliability, provided that the Markov chain simulation yields samples that populate a sufficient part of the failure domain. In the case where FORM analysis is performed for multiple design points, the initial samples can also be generated using the stratified density of Eq. (4.92).

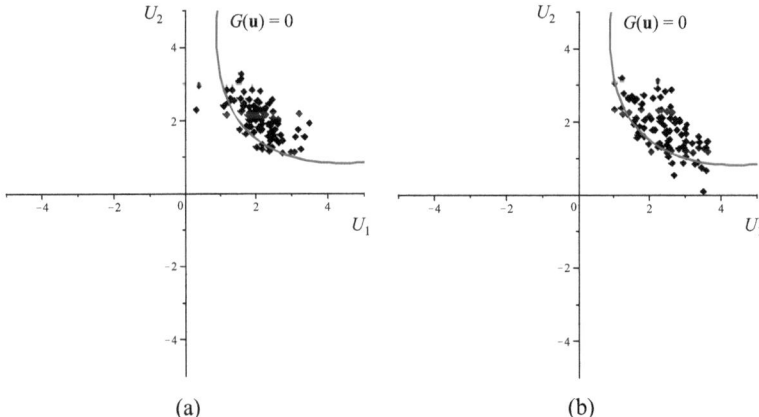

(a)   (b)

Figure 4.14: Graphical representation of adaptive importance sampling methods in a 2D standard normal space (100 samples). (a) Parametric adaptive sampling. (b) Importance sampling using kernels.

Adaptive directional importance sampling

An adaptive directional importance sampling procedure has been proposed by the author [96]. The procedure is performed in two steps. First, an initial coarse number of deterministic, optimally distributed directions are generated, utilizing one of the algorithms discussed in Section 4.7.3, and the corresponding radii $\{r_i = r(\mathbf{a}_i), (i = 1,\ldots,m)\}$ are computed. Then an adaptive sampling density can be computed as the following stratified sampling density:

$$h_a(\mathbf{a}) = \frac{\sum_{i=1}^{m} \Phi(-r_i) h_i(\mathbf{a})}{\sum_{i=1}^{m} \Phi(-r_i)} \qquad (4.112)$$

Where m is the number of uniformly distributed directions and $h_i(\mathbf{a})$ is the importance sampling density corresponding to the radius r_i, given by:

$$h_i(\mathbf{a}) = \frac{1 - \chi_n^2\left[\left(\dfrac{r_i}{\mathbf{a}^T\mathbf{a}_i}\right)^2\right]}{\Phi[-r_i]} f_A(\mathbf{a}) \qquad \mathbf{a} \in \{D_A | \mathbf{a}^T \mathbf{a}_i > 0\} \qquad (4.113)$$

where $f_A(\mathbf{a})$ is the uniform PDF on the n-dimensional hypersphere D_A. This method has the advantage that the design point is not computed, while the equally distributed initial directions provide a good estimation of the optimal importance sampling density. However, the method becomes inefficient with increasing dimension n, since the required number of initial directions for the estimation of the optimal sampling density increases considerably. The initial estimate of the probability computed using the deterministic directions can be combined with the results obtained by the importance sampling density, applying the following mixing sampling density:

$$h_m(\mathbf{a}) = p \cdot f_A(\mathbf{a}) + (1-p) \cdot h_a(\mathbf{a}) \qquad (4.114)$$

where $p = m/(m + N)$. The estimate of the probability of failure reads:

$$\hat{P}_f = \frac{p}{m}\sum_{i=1}^{m}\left[1-\chi_n^2\left(r_i^2\right)\right] + \frac{1-p}{N}\sum_{i=1}^{N}\left[1-\chi_n^2\left(r(\mathbf{a}_i)^2\right)\right]\frac{f_\mathbf{A}(\mathbf{a}_i)}{h_a(\mathbf{a}_i)} \qquad (4.115)$$

where the samples $\{\mathbf{a}_k, (k = 1,...,N)\}$ are simulated according to $h_a(\mathbf{a})$.

4.7.5 Comparison of the simulation methods

In this section, the simulation method presented in Sections 4.7.2-4.7.4 are illustrated for the reliability analysis of the limit-state function of Eq. (4.72), wherein the random variables are independent and standard normal. For comparison, the results obtained by the FORM are $\beta = 2.5$, $P_f = 6.209 \times 10^{-3}$, and the asymptotic results by the SORM $\beta = 2.62$, $P_f = 4.39 \times 10^{-3}$. For the methods using the design point, the point $\mathbf{u}^*$ obtained by the iHL-RF method is $\mathbf{u}^* = [1.7678\ 1.7678]^T$. Also, a standard importance sampling simulation based on the design point with a sample size of $N = 10^5$ gave an estimate of the probability of failure $\hat{P}_f = 4.21 \cdot 10^{-3}$ with a coefficient of variation smaller than 0.1%.

Figure 4.15 shows the estimated probability of failure obtained by the Monte Carlo (MC) method and the coefficient of variation of the estimate versus the number of samples, with every point corresponding to an independent simulation run. The three cases considered are based on the method applied for the generation of the samples, i.e. pseudo-random sampling (MC), Latin hypercube sampling (LHS) and quasi-random sampling (QRS) with the Niederreiter sequence points. The results show an improvement of the Monte Carlo method when more elaborate sampling techniques such as LHS and QRS are used, with the QRS shown to be the most efficient of the two. This is better illustrated in Figure 4.15(a), wherein it is shown that the QRS reaches a stable solution with already 4000 samples, whereas the MC and LHS still present significant oscillations in the neighborhood of the exact solution. Similar results were obtained in [17] for several example cases, wherein it is concluded that QRS methods have faster rate of convergence than MC and LHS. In practical terms, QRS techniques can be applied with larger tolerances in the

132   4 Finite element reliability assessment

coefficient of variation of the estimate of the failure probability, as compared to the ones used in standard Monte Carlo methods.

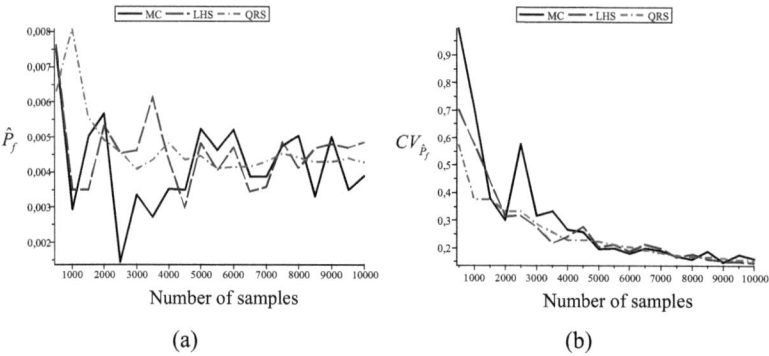

(a)   (b)

Figure 4.15: Estimate of P_f (a) of the example and coefficient of variation of the estimate (b) in terms of the number of samples for the Monte Carlo with pseudo-random sampling (MC), the Latin hypercube sampling (LHS) and the quasi-random sampling (QRS) with the Niederreiter sequence points.

In Figure 4.16, the results obtained by the directional simulation method are compared for the cases where the points (directions) are generated randomly (RP) and deterministically (DP) by application of the geometric method, discussed in Section 4.7.3. Again each point in the figure corresponds to an independent simulation. It is shown that a remarkable improvement of the performance of the directional simulation is obtained when an optimally distributed set of directions is used. For this case, the exact value of the failure probability is obtained with only 20 directions, whereas the standard directional simulation with random directions presents a strong oscillatory behavior and requires a much larger number of samples to reach the same level of accuracy. The benefit of the use of deterministic directions in the directional simulation method is illustrated for several analytical and numerical limit-states in [65], [86] and [95].

4.7 Simulation methods 133

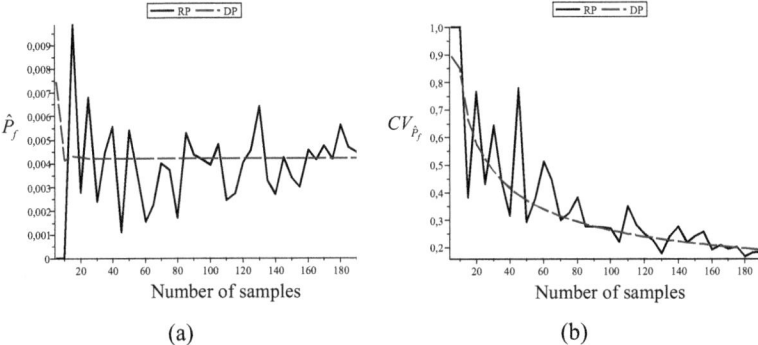

Figure 4.16: Estimate of P_f (a) of the example and coefficient of variation of the estimate (b) in terms of the number of samples for the directional simulation with random points (directions) (RP), and with deterministic points (DP) generated by the geometric method.

In Figure 4.17, the standard importance sampling (SIS), conditional sampling (CS) and conditional importance sampling (CIS) methods are compared. In is shown that for this example, where the limit-state function is quadratic at the design point, the SIS performs better than the CS, i.e. for the same number of samples SIS gives better estimates and smaller coefficients of variation as compared to CS. Moreover, the CIS, which combines the advantages of the other two methods, presents the best behavior in terms of stability [Figure 4.17(a)] and gives the smallest coefficients of variation [Figure 4.17(b)]. It should be noted that in the case of multiple design points or hyperspherical limit-states, the CS is expected to perform better than the SIS [71]. In the case where all design points are found, then a stratified importance sampling function based on the CIS method could be applied.

The axis orthogonal importance sampling (AOIS) and directional importance sampling (DIS) methods are compared in Figure 4.18. It is shown that the two methods have similar performances. Both methods require the solution of a line-search problem for each sample point. Therefore, the computational cost depends on the number of limit-state function evaluation needed to solve the corresponding line-search. In this

134 4 Finite element reliability assessment

study, the secant method is used and for this example an average of 5 limit-state function evaluations per line-search were needed for both methods.

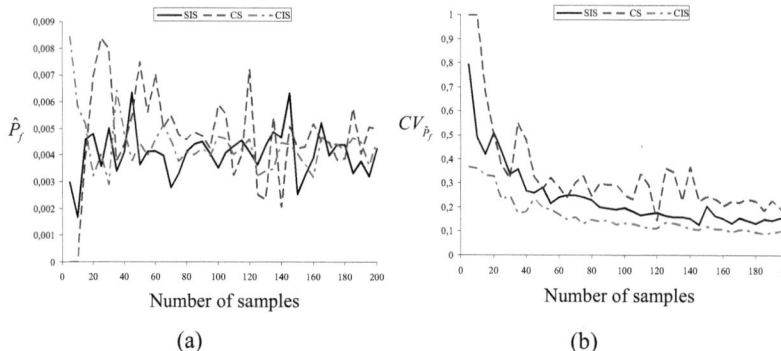

(a)   (b)

Figure 4.17: Estimate of P_f (a) of the example and coefficient of variation of the estimate (b) in terms of the number of samples for the standard importance sampling (SIS), the conditional sampling (CS) and conditional importance sampling (CIS).

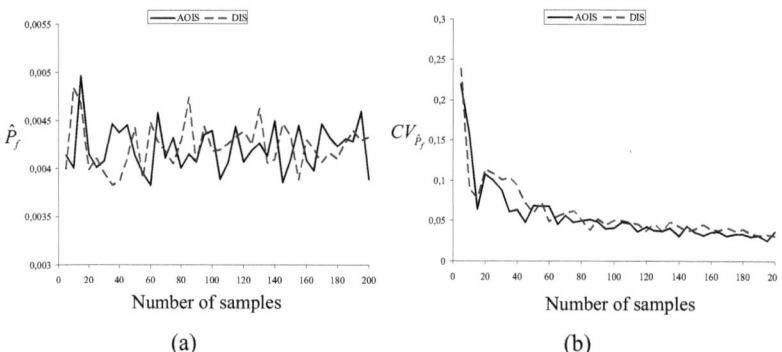

(a)   (b)

Figure 4.18: Estimate of P_f (a) of the example and coefficient of variation of the estimate (b) in terms of the number of samples for the axis orthogonal importance sampling (AOIS) and the directional importance sampling (DIS).

In Figure 4.19, the results obtained by the adaptive importance sampling (AIS) method with the joint normal sampling function of Eq. (4.108) are shown. The mean vector and covariance matrix entering the definition of the sampling function were computed using 100 samples generated by the SIS method. Comparing Figure 4.19 with Figure 4.17, one can see that AIS shows a significant improvement of the probability estimates and the coefficient of variation obtained by SIS. For instance, a coefficient of variation of 20% is obtained by SIS with 200 samples, while AIS with a total of 200 samples (including the samples of the initial simulation for the determination of the adaptive sampling function) leads to a coefficient of variation smaller than 10%.

Finally, Figure 4.20 presents the results obtained by the adaptive directional importance sampling (ADIS) method. The radii entering the definition of the adaptive sampling function of Eq. (4.113) are determined by an initial simulation with 50 deterministic directions. Comparing Figure 4.20 with Figure 4.18, we see that ADIS has a comparable performance to the one of the DIS. However, the advantage of the ADIS is that the design point is not needed. Therefore this method can be applied to cases where the limit-state function has multiple design points, or for highly nonlinear implicit limit-states for which the convergence of the FORM algorithms is not guaranteed.

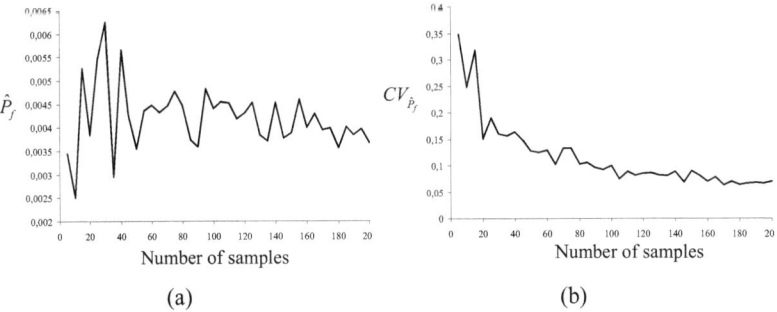

Figure 4.19: Estimate of P_f (a) of the example and coefficient of variation of the estimate (b) in terms of the number of samples for adaptive importance sampling, based on 100 samples simulated by standard importance sampling.

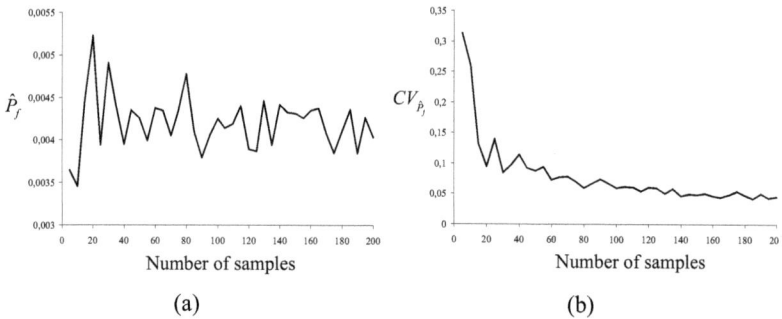

Figure 4.20: Estimate of P_f (a) of the example and coefficient of variation of the estimate (b) in terms of the number of samples for the adaptive directional importance sampling, based on 50 deterministic directions generated by the geometric method.

4.8 Simulation in high dimensions

This section discusses the case where a large number of basic random variables (> 50) are included in the definition of the limit-state function. This is often the case when the time or spatial variability of an uncertain quantity requires the discretization of the corresponding random process or field. For such problems, the evaluation of the design point becomes inefficient, since the computational cost for the numerical evaluation of the derivatives in the FORM optimization algorithm is a function of the number of random variables. This problem can be solved if the derivatives are evaluated applying the direct differentiation method ([26], [142]). However, this method needs alterations at the FE code level and hence cannot be used in conjunction with a "black-box" FE code. Therefore, the importance sampling methods presented in Section 4.7.4 become inefficient for such problems. To circumvent this problem, a number of simulation methods have been specially developed for the treatment of such high-dimensional problems. These include the subset simulation [6], the spherical subset simulation [64] and the asymptotic sampling method [13]. In the following, the subset simulation is discussed in detail.

4.8.1 The subset simulation

The subset simulation is an adaptive simulation method developed by Au and Beck [6]. The method is based on the standard Monte Carlo simulation but overcomes its inefficiency in estimating small probabilities, while maintaining its independency on the problem dimensionality. This is achieved by expressing the failure event $F = \{G(\mathbf{U}) \leq 0\}$ as the intersection of M intermediate failure events:

$$F = \bigcap_{i=1}^{M} F_i \qquad (4.116)$$

where $F_1 \supset F_2 \supset \cdots \supset F_M = F$. The probability of failure is estimated by computing the joint probability $P(\bigcap_{i=1}^{M} F_i)$ as a product of conditional probabilities:

$$P_f = P(F) = P\left(\bigcap_{i=1}^{M} F_i\right) = P(F_1) \prod_{i=2}^{M} P(F_i | F_{i-1}) \qquad (4.117)$$

The failure event is defined in the equivalent standard normal space by $G(\mathbf{u}) \leq 0$; each of the intermediate events is defined as $F_i = \{G(\mathbf{u}) \leq G_i\}$, where $G_i > \ldots > G_M = 0$. The values of G_i are chosen adaptively so that the estimates of the conditional probabilities correspond to a chosen value p_0. The probability of F_1 is computed by applying the crude Monte Carlo method. Through a Monte Carlo simulation, N samples of $\mathbf{U}$ are simulated and G_1 is set equal to the $[(1 - p_0)N]$-th largest value among the samples $\{G(\mathbf{u}_k): k = 1,\ldots,N\}$. The $p_0 N$ samples $\mathbf{u}_i$ for which $G(\mathbf{u}_i) \leq G_1$ are used as starting points for the simulation of $(1 - p_0)N$ samples conditional on F_1, by applying a modified version [6] of the Metropolis-Hastings algorithm ([54], [84]), which is a Markov Chain Monte Carlo (MCMC) technique. This procedure is repeated for sampling conditional on $F_2, F_3,\ldots$ until the maximum level M is reached, for which the threshold $G_M = 0$ is given. An estimate of the failure probability is then given by:

$$\hat{P}_f = p_0^{M-1} \hat{P}(F_M | F_{M-1}) \qquad (4.118)$$

where the estimate $\hat{P}(F_M | F_{M-1})$ of the conditional probability is as follows:

$$\hat{P}(F_M | F_{M-1}) = \frac{1}{N}\sum_{k=1}^{N} I(\mathbf{u}_k) \qquad (4.119)$$

where $I(\mathbf{u})$ is the indicator function defined in Eq. (4.67) and $\{\mathbf{u}_k \ (k = 1,\ldots,m)\}$ are simulated conditional on F_{M-1}. It is shown in [6] that the estimator $\hat{P}_f$ is biased for a finite N, due to the correlation between the estimates of the conditional probability, but it is asymptotically unbiased.

The analyst is free in the choice of p_0 and the number of samples at each step N. However, N should be selected large enough to give an accurate estimate of p_0. If the magnitude of P_f is in the order of 10^{-k} then the total required number of samples is $N_{tot} = k(1 - p_0)N + N$. Note that for the crude Monte Carlo, the required number of samples for the same probability and a target coefficient of variation of 0.1 is 10^{k+2}, which indicates that the gain in efficiency of the subset simulation can be of several orders of magnitude. A graphical representation of this method, applied for the reliability analysis of the limit-state function of Eq. (4.72), is shown in Figure 4.21, for $p_0 = 0.1$ and $N = 400$.

Modified Metropolis-Hastings algorithm

Markov chains are random processes, defined such that every state (time instant) depends only on the previous state and not on the states that preceded it. In other words, the random variable at state i is distributed according to a conditional PDF, given the variable at state $i - 1$. The MCMC methods generate Markov chain samples which are asymptotically distributed according to a given distribution [42]. In the subset simulation, the conditional samples on the intermediate failure domains $\{F_i \ (i = 2,\ldots,M)\}$ are computed through MCMC, by application of a modified version [6] of the Metropolis-Hastings algorithm.

4.8 Simulation in high dimensions

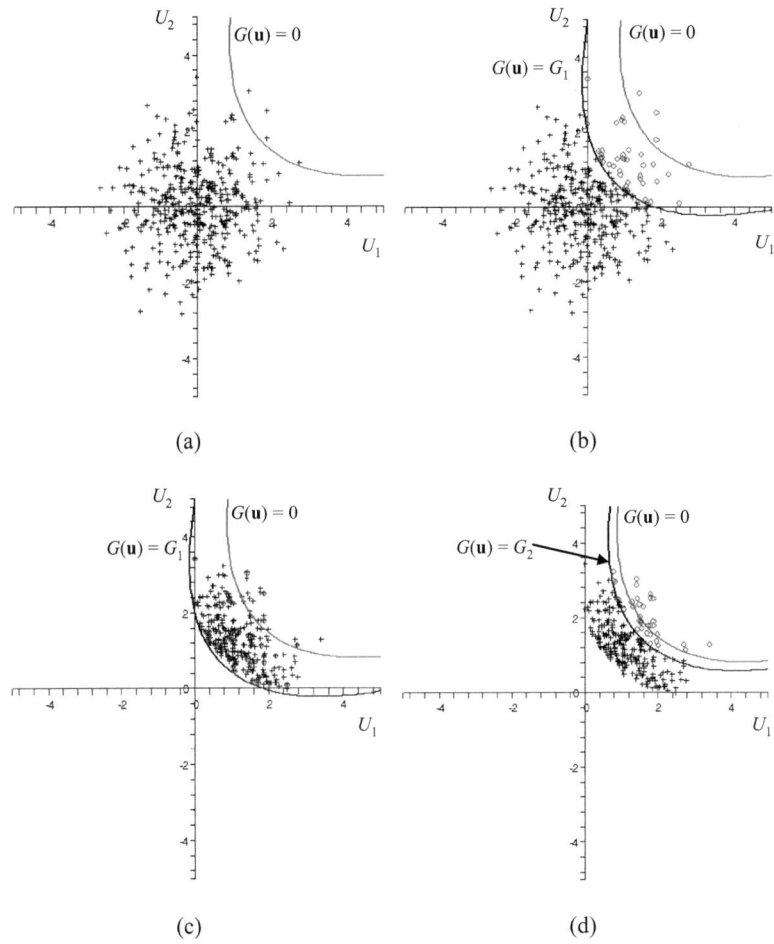

Figure 4.21: Graphical representation of the subset simulation in a 2D standard normal space (a) Monte Carlo simulation. (b) Determination of the first threshold G_1. (c) Markov chain simulation of the intermediate failure domain. (d) Determination of the second threshold G_2.

Assume that at the level i, the samples $\{\mathbf{u}_k \ (k = 1,\ldots,m)\}$ fall in the failure domain F_{i+1}. To simulate samples of $\mathbf{U}$ conditional on F_{i+1} we apply the following procedure. Let $f^*(v|u_{kj})$ be a one-dimensional conditional PDF centered at u_{kj} with the symmetry property $f^*(v|u_{kj}) = f^*(u_{kj}|v)$, where u_{kj} is

140   4 Finite element reliability assessment

the j-th component of $\mathbf{u}_k$. For every $\mathbf{u}_k$, $k = 1,\ldots,m$, we generate the sequence of samples $\{\mathbf{v}_l \ (l = 1,\ldots,N/m)\}$, which corresponds to one chain, starting from a given $\mathbf{v}_1 = \mathbf{u}_k$ (the seed of the chain) by computing $\mathbf{v}_{l+1}$ from $\mathbf{v}_l$, $l = 1,\ldots, N/m - 1$, as follows:

1. Generate a candidate state $\tilde{\mathbf{v}}$: For each component $\tilde{v}_j$, $j = 1,\ldots,n$ of $\tilde{\mathbf{v}}$ simulate ξ_j from $f^*(\xi_j|v_{lj})$. Compute the ratio $r_j = \varphi(\xi_j)/\varphi(v_{lj})$. Then set $\tilde{v}_j = \xi_j$ with probability $\min\{1, r_j\}$ and $\tilde{v}_j = v_{lj}$ with probability $1 - \min\{1, r_j\}$.

2. Choose the next state $\mathbf{v}_{l+1}$: Compute $G(\tilde{\mathbf{v}})$. If $\tilde{\mathbf{v}} \in F_{i+1}$ set $\mathbf{v}_{l+1} = \tilde{\mathbf{v}}$, otherwise set $\mathbf{v}_{l+1} = \mathbf{v}_l$.

It is shown in [6] that each new sample $\mathbf{v}_{l+1}$ will be distributed according to $\varphi_n(.|F_{i+1})$ if $\mathbf{v}_l$ also is, where:

$$\varphi_n(\mathbf{u}|F_{i+1}) = \frac{\varphi_n(\mathbf{u})}{P(F_{i+1})} \qquad \mathbf{u} \in F_{i+1} \qquad (4.120)$$

is the stationary PDF of the Markov chain. The efficiency of the algorithm is shown to be insensitive to the type of the PDF $f^*(v|u_{kj})$. In this study, the uniform PDF centered at each sample u_{kj} is used, as suggested in [6]. The width of the PDF $f^*(v|u_{kj})$ influences the size of F_{i+1} covered by the Markov chain samples. A large width may decrease the rate of acceptance of the samples, but guarantees ergodicity of the Markov chain, i.e. it assures that the stationary distribution $\varphi_n(.|F_{i+1})$ is unique and independent of the initial state – although it should be noted that since the Markov chain samples are simulated starting from each point $\mathbf{u}_k$, $k = 1,\ldots,m$, the assumption of ergodicity is usually valid. On the other hand, a very small width increases the dependence between the samples, which may influence the quality of the estimator and can also counteract the assumption of ergodicity. A compromise width of 2, i.e. twice the standard deviation at the U-space, is used in this study.

Based on the MCMC procedure we can derive expressions for the variance of the estimators of the conditional probabilities. For example, the variance of the estimator $\hat{P}_M = \hat{P}(F_M | F_{M-1})$ in Eq. (4.119) is given as follows [6]:

$$\text{Var}(\hat{P}_M) = \frac{P_M(1-P_M)}{N}(1+\gamma) \qquad (4.121)$$

where

$$\gamma = 2 \sum_{k=1}^{N/m-1} \left(1 - \frac{km}{N}\right) \frac{R(k)}{R(0)} \qquad (4.122)$$

and

$$R(k) = \text{E}\left[I(\mathbf{u}_{j,l})I(\mathbf{u}_{j,l+k})\right] - P_M^2 \qquad (4.123)$$

where m is the number of seeds for the MCMC, N/m is the length of each chain, $\mathbf{u}_{j,l}$ is the l-th sample of the j-th chain and $R(k)$ is the covariance between $I(\mathbf{u}_{j,l})$ and $I(\mathbf{u}_{j,l+k})$, which is independent of j and l, since all chains are stationary and equivalent. Also, it has been assumed that the seeds for each chain are uncorrelated through the indicator function $I(\mathbf{u})$, i.e. that $\text{E}\left[I(\mathbf{u}_j)I(\mathbf{u}_k)\right] - P_M^2 = 0$ if $\mathbf{u}_j$ and $\mathbf{u}_k$ belong to different chains. Comparing Eq. (4.121) and Eq. (4.70) we see that the variance of the MCMC estimator is larger than the one of the crude Monte Carlo method. This is due to the dependence of the samples of the Markov chain (i.e. $\gamma > 0$).

4.9 Response surface methods

In the previous sections it is shown that a large number of numerical evaluations of the limit-state function (i.e. calls to the FE solver) may be required for an accurate estimation of the probability of failure. Response surface methods are based on approximating the limit-state function using a simple mathematical model [38]. Then the reliability analysis can be performed using an analytical expression instead of the true limit-state function. This approach may reduce the computational effort considerably, provided that a sufficiently accurate approximation of the limit-state function is built with a limited number of calls to the FE solver.

Let $\hat{g}(\mathbf{X})$ be the approximation of the limit-state function $g(\mathbf{X})$ in the basic random variable space $\mathbf{X}$. Typically $\hat{g}(\mathbf{X})$ is of quadratic polynomial form. In this study, two different models are used. The first is a quadratic polynomial without mixed terms:

$$\hat{g}(\mathbf{x}) = c_0 + \sum_{i=1}^{n} c_i x_i + \sum_{i=1}^{n} c_{ii} x_i^2 \qquad (4.124)$$

where the $(1 + 2n)$ coefficients $\mathbf{c} = [c_0, \{c_i, i = 1,\ldots,n\}, \{c_{ii}, i = 1,\ldots,n\}]^T$ are to be determined. Eq. (4.124) can be enriched by adding the mixed terms and therefore accounting for possible interaction between the random variables:

$$\hat{g}(\mathbf{x}) = c_0 + \sum_{i=1}^{n} c_i x_i + \sum_{i=1}^{n} c_{ii} x_i^2 + \sum_{i=1}^{n} \sum_{j=i+1}^{n} c_{ij} x_i x_j \qquad (4.125)$$

where in this case the total number of unknown coefficients is $[1 + n + n(n - 1)/2]$ and $\mathbf{c} = [c_0, \{c_i, i = 1,\ldots,n\}, \{c_{ii}, i = 1,\ldots,n\}, \{c_{ji}, i = 1,\ldots,n, j = i+1,\ldots,n\}]^T$.

The unknown coefficients $\mathbf{c}$ are determined by the least squares method. First, a set of experimental design points $\{\mathbf{x}^j, j = 1,\ldots,K\}$ are chosen, for which the exact value of the limit-state function $y^j = g(\mathbf{x}^j)$ is computed. The coefficients $\mathbf{c}$ are found by requiring the sum of squares of the differences between the value of the function $\hat{g}(\mathbf{x}^j)$ and the computed actual value y^j at the K experimental points to be minimum, i.e.

$$\mathbf{c} = \arg\min \left\{ \sum_{j=1}^{K} \left(y^j - \hat{g}(\mathbf{x}^j; \mathbf{c}) \right)^2 \right\} \qquad (4.126)$$

Let $\mathbf{q}(\mathbf{x})$ be a vector of dimension n_c whose entries are the monomials included in $\hat{g}(\mathbf{x})$ of either Eq. (4.124) or Eq. (4.125), where n_c is the number of unknown coefficients. Then the function $\hat{g}(\mathbf{x})$ can be expressed as follows:

4.9 Response surface methods

$$\hat{g}(\mathbf{x}) = \mathbf{q}(\mathbf{x})^T \mathbf{c} \qquad (4.127)$$

Using Eq. (4.127), the solution of the problem of Eq. (4.126) is as follows (e.g. see [38]):

$$\mathbf{c} = \left(\mathbf{Q}^T \mathbf{Q}\right)^{-1} \mathbf{Q}^T \mathbf{y} \qquad (4.128)$$

where $\mathbf{Q}$ is a $K \times n_c$ matrix whose rows are the vectors $\mathbf{q}(\mathbf{x}^j)^T$ and $\mathbf{y}$ is a vector with components $y^j = g(\mathbf{x}^j)$. The solution of Eq. (4.128) requires $K \geq n_c$.

Several different experimental designs have been proposed for the selection of the set $\{\mathbf{x}^j, j = 1,\ldots,K\}$. The considered designs consist of a grid of points centered around the mean value vector $\boldsymbol{\mu}_\mathbf{X}$. For the model of Eq. (4.124), an axial design has been proposed [14], where the central point is taken as the mean value and two additional points are taken at each Cartesian axis (see Figure 4.22), i.e.

$$\begin{aligned} \mathbf{x}^1 &= \boldsymbol{\mu}_\mathbf{X} \\ \mathbf{x}^j &= \boldsymbol{\mu}_\mathbf{X} \pm k\sigma_{X_i} \mathbf{e}_i \qquad \forall i: 1 \leq i \leq n \end{aligned} \qquad (4.129)$$

where $\mathbf{e}_i$ is the i-th unit Cartesian vector and $k \in \mathbb{R}_+$. The axial design leads to $K = 1 + 2n$ points, which coincides with the number of unknowns n_c for the model of Eq. (4.124). The quality of the approximation is shown to be strongly dependent on the choice of k [50]. For most structural reliability problems, $k = 3$ seems to be an appropriate choice.

For the model of Eq. (4.125) a choice of two different designs is provided. The first is the full factorial design, which generates l points for each coordinate and includes all possible combinations thus producing a total of $K = l^n$ points. For $l = 3$ the points at each Cartesian axis can be constructed as follows:

$$\begin{aligned} x_i^1 &= \mu_{X_i} \\ x_i^j &= \mu_{X_i} \pm k\sigma_{X_i} \end{aligned} \qquad (4.130)$$

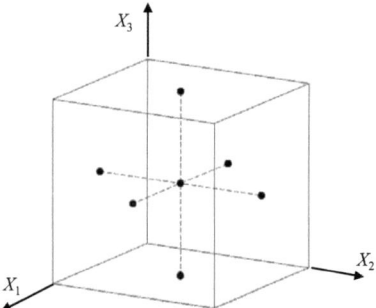

Figure 4.22: Axial design for $n = 3$.

A full factorial design for $l = 3$ and $n = 3$ is shown in Figure 4.23(a). Alternatively, the less expensive central composite design can be applied. This design combines a full factorial design for $l = 2$ and the axial design of Eq. (4.129), leading to $K = 1 + 2n + 2^n$ points [38] [see Figure 4.23(b)].

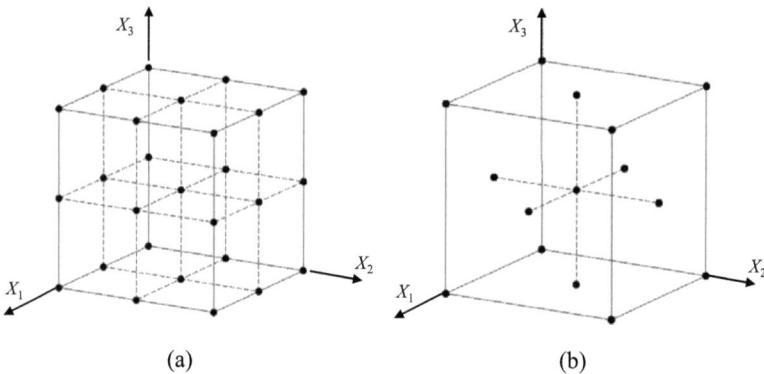

Figure 4.23: (a) Full factorial design for $n = 3$. (b) Central composite design for $n = 3$.

To enhance the quality of the approximation, an adaptive procedure can be employed in combination with any of the response surface models and experimental designs. This procedure opts for a better approximation at the area close to the design point. In the context of the FORM optimization, the response surface is adapted iteratively by centering the experimental design at a point $\mathbf{x}_0$, obtained by a linear interpolation between the design point $\mathbf{x}^*$ and $\boldsymbol{\mu}_\mathbf{X}$, so that $g(\mathbf{x}_0) \approx 0$, i.e.:

$$\mathbf{x}_0 = \boldsymbol{\mu}_\mathbf{X} + \left(\mathbf{x}^* - \boldsymbol{\mu}_\mathbf{X}\right)\frac{g(\boldsymbol{\mu}_\mathbf{X})}{g(\boldsymbol{\mu}_\mathbf{X}) - g(\mathbf{x}^*)} \qquad (4.131)$$

Then a new FORM optimization is performed and the procedure is repeated until the difference between two subsequent reliability indices is smaller than a prescribed tolerance (see [14] and [108]).

Several other models have been proposed in the literature for approximating the limit-state function. In [103], the moving least squares method is applied to construct a local approximation of the limit-state function close to the design point. Also several researchers have used artificial neural networks to model the response surface (e.g. see [15], [19], [62], [92], [120]). It is shown that the efficiency of this approach depends on a properly chosen training set. Finally, another method benefits from the fact that the Monte Carlo method only requires the knowledge on whether failure has occurred for each sample and thus approaches the problems from the perspective of data classification ([61], [109]). This approach uses support vector machines to built a pattern recognition scheme that only indicates whether failure has occurred.

5 Bayesian updating of reliability estimates

It is common in engineering applications that additional information about the structure appears at states after the conclusion of the design procedure. This information can be related to inspection, instrumented monitoring, knowledge on the exceedance of a failure condition or historically survived loads. This chapter describes how this information can be used to update the estimate of the reliability of the structure. First, the basic concepts are introduced. Then, methods that have been developed for dealing with a special category of reliability updating problems are discussed in detail.

5.1 Introduction

In the basic variable space $\mathbf{X}$, we can express any available information about the structure by using a limit-state function $h(\mathbf{x})$. The information can be of two different types; the inequality-type information is defined by an inequality of the form $\{h(\mathbf{x}) \leq 0\}$ and usually refers to information about an exceedance of some limit state; on the other hand, the equality-type information is defined by an equality, i.e. $\{h(\mathbf{x}) = 0\}$, and usually refers to some measurement outcome. Each information can then be expressed by an event H, defined as $H = \{h(\mathbf{x}) \leq 0\}$ or $H = \{h(\mathbf{x}) = 0\}$. Assuming that m

such events are available, each of which is denoted by H_i, the event $H = H_1 \cap \ldots \cap H_m$ will contain all available information. Denoting by F the failure event of the structure defined as $F = \{g(\mathbf{x}) \leq 0\}$, where $g(\mathbf{x})$ is the generalized limit-state function [see Eq. (3.22)], we can express the probability of failure conditional on the information event H, as follows:

$$P(F|H) = \frac{P(F \cap H)}{P(H)} = \frac{\int_{F \cap H_1 \cap \ldots \cap H_m} f_\mathbf{X}(\mathbf{x}) d\mathbf{x}}{\int_{H_1 \cap \ldots \cap H_m} f_\mathbf{X}(\mathbf{x}) d\mathbf{x}} \qquad (5.1)$$

where $f_\mathbf{X}(\mathbf{x})$ is the joint PDF of $\mathbf{X}$. If all information events H_i are of inequality-type, then the computation of the integrals in Eq. (5.1) is straightforward and can be done by application of any of the reliability methods described in Chapter 4. However, if one or more events H_i are of equality-type then the domain of integration for both integrals in Eq. (5.1) reduces to a surface consisting of a subset of all points for which $h_i(\mathbf{x}) = 0$ at the n-dimensional basic variable space and direct application of reliability methods is not possible. In practical terms, the probability of an equality-type event H_i is zero, since the random variable $h_i(\mathbf{X})$ is continuous, and therefore both integrals in Eq. (5.1) will result in zero. To circumvent this problem, several approaches have been suggested, which will be discussed in the following sections.

5.2 Updating with equality information

In the following, it is assumed for convenience that all information events H_i are of equality-type. The solution of problems that include both types of information is a straightforward extension.

5.2.1 First- and second-order methods

This section discusses the first- and second-order approximation concepts developed for the solution of reliability updating problems with equality

5.2 Updating with equality information

information. The method presented in [79] suggests to include a set of dummy parameters $\mathbf{d} = [d_1, \ldots, d_m]^T$ and express the conditional probability of Eq. (5.1) as follows:

$$P(F | H_1 \cap \ldots \cap H_m) = \frac{\left.\dfrac{\partial^m P\big(g(\mathbf{x}) \leq 0 \cap h_1(\mathbf{x}) - d_1 \leq 0 \cap \ldots \cap h_m(\mathbf{x}) - d_m \leq 0\big)}{\partial d_1 \ldots \partial d_m}\right|_{d=0}}{\left.\dfrac{\partial^m P\big(h_1(\mathbf{x}) - d_1 \leq 0 \cap \ldots \cap h_m(\mathbf{x}) - d_m \leq 0\big)}{\partial d_1 \ldots \partial d_m}\right|_{d=0}} \quad (5.2)$$

In the expression of Eq. (5.2), the equality-type events $\{h_i(\mathbf{x}) = 0\}$ are replaced by the inequality-type events $\{h_i(\mathbf{x}) - d_i \leq 0\}$. This approach requires the computation of the partial derivatives of the probabilities with respect to the dummy parameters $\mathbf{d}$, evaluated at $\mathbf{d} = \mathbf{0}$. This can be done by using the FORM (or SORM) results for parallel system reliability problems. The FORM approximation of the probability in the nominator of Eq. (5.2) is as follows:

$$P\big(g(\mathbf{x}) \leq 0 \cap h_1(\mathbf{x}) - d_1 \leq 0 \cap \ldots \cap h_m(\mathbf{x}) - d_m \leq 0\big) \approx \Phi_{m+1}(-\mathbf{B}_{m+1}, \mathbf{R}_{m+1}) \quad (5.3)$$

where $\Phi_{m+1}(.)$ is the $(m + 1)$-variate standard normal CDF, $\mathbf{B}_{m+1}$ contains the FORM reliability indices of the $m + 1$ component reliability problems and $\mathbf{R}_{m+1}$ is the correlation matrix of the random variables $Y_i = \mathbf{a}_i^T \mathbf{U}$ with entries $\rho_{ij} = \mathbf{a}_i^T \mathbf{a}_j$, $\mathbf{U}$ being the equivalent standard normal space (see Section 4.6). The mixed partial derivative of Eq. (5.3) with respect to the parameters $\mathbf{d}$ can be computed by the following expression [79]:

$$\left.\frac{\partial^m P\big(g(\mathbf{x}) \leq 0 \cap h_1(\mathbf{x}) - d_1 \leq 0 \cap \ldots \cap h_m(\mathbf{x}) - d_m \leq 0\big)}{\partial d_1 \ldots \partial d_m}\right|_{d=0}$$
$$\approx \frac{\partial^m \Phi_{m+1}(-\mathbf{B}_{m+1}, \mathbf{R}_{m+1})}{\partial \beta_1 \ldots \partial \beta_m} \quad (5.4)$$
$$= \varphi_m(-\mathbf{B}_m, \mathbf{R}_m) \Phi\left(-\big(\beta - \boldsymbol{\rho}_m^T \mathbf{R}_m^{-1} \mathbf{B}_m\big), 1 - \boldsymbol{\rho}_m^T \mathbf{R}_m^{-1} \boldsymbol{\rho}_m\right)$$

where $\mathbf{B}_m$ contains the reliability indices β_i, $\mathbf{R}_m$ is the correlation matrix of the variables $Y_i = \boldsymbol{\alpha}_i^T \mathbf{U}$, with β_i and $\boldsymbol{\alpha}_i$ being the component reliability index and vector of influence coefficients for the event $\{h_i(\mathbf{x}) - d_i \leq 0\}$ evaluated at $d_i = 0$, $\boldsymbol{\rho}_m$ is an m-dimensional vector with entries $\rho_i = \boldsymbol{\alpha}^T \boldsymbol{\alpha}_i$, with β and $\boldsymbol{\alpha}$ being the reliability index and vector of influence coefficients of the original reliability problem, i.e. for the event $\{g(\mathbf{x}) \leq 0\}$, and $\varphi_m(.)$ is the m-variate standard normal PDF. Also it has been used that $\partial \beta_i / \partial d_i = -1$, $\forall\ i$: $1 \leq i \leq m$. A similar expression can also be derived for the denominator in Eq. (5.2). In the case where the original reliability problem is a parallel or series system problem, the expression in Eq. (5.4) should be modified accordingly. The same results can also be obtained by a different approach that involves first- or second-order approximations of surface integration of multi-normal densities [116].

5.2.2 A general approach

This section discusses a general approach [130] for dealing with reliability updating problems with equality information, which allows the application of any reliability method, including simulation methods. Since the information events H_i usually refer to measurement outcomes, we can express the corresponding limit-state functions, as follows:

$$h_i(\mathbf{x}) = \hat{h}_i(\mathbf{x}_{-i}) + x_i \qquad (5.5)$$

where X_i is a random variable representing the measurement error and $\mathbf{X}_{-i}$ denotes the set of all random variables $\mathbf{X}$ excluding X_i. If a structural system characteristic $s(\mathbf{x}_{-i})$ is measured, e.g. the deformation of a structural member, then the function $\hat{h}_i(\mathbf{x}_{-i})$ can be expressed as $\hat{h}_i(\mathbf{x}_{-i}) = s(\mathbf{x}_{-i}) - s_m$, where s_m is the measurement outcome. The information of the event H_i with respect to the random variables $\mathbf{X}_{-i}$ can also be expressed by the likelihood function [2]:

$$L_i(\mathbf{x}_{-i}) \propto P(H_i | \mathbf{X}_{-i} = \mathbf{x}_{-i}) = f_{X_i}\left[-\hat{h}_i(\mathbf{x}_{-i})\right] \qquad (5.6)$$

5.2 Updating with equality information

where $f_{X_i}(.)$ is the PDF of X_i. Eq. (5.6) assumes that X_i is statistically independent of $\mathbf{X}_{-i}$. If this is not the case, independence can be achieved using the transformation methods discussed in Section 4.2. Also, if the equality limit-state functions cannot be expressed in the form of Eq. (5.5), then the likelihood function will take the following form:

$$L_i(\mathbf{x}_{-i}) \propto \sum_{j=1}^{n_i} f_{X_i}\left[\hat{x}_{ij}(\mathbf{x}_{-i})\right] \tag{5.7}$$

where $\hat{x}_{ij}(\mathbf{x}_{-i})$ are the n_i roots of the equation $h_i(\mathbf{x}_{-i}, x_i) = 0$.

An equivalent formulation of the likelihood function is given by:

$$L_i(\mathbf{x}_{-i}) = \frac{1}{c_i}\Pr\left\{U_i - \Phi^{-1}\left[c_i L_i(\mathbf{x}_{-i})\right] \leq 0\right\} \tag{5.8}$$

where U_i is a standard normal random variable, $\Phi^{-1}(.)$ is the inverse of the standard normal CDF and c_i is a positive constant, chosen to ensure that $c_i L_i(\mathbf{x}_{-i}) \leq 1$ for all $\mathbf{x}_{-i}$. Eq. (5.8) enables the expression of the likelihood function by an equivalent inequality information event $H_{e,i} = \{h_{e,i}(\mathbf{x}_{-i}, u_i) \leq 0\}$, with:

$$h_{e,i}(\mathbf{x}_{-i}, u_i) = u_i - \Phi^{-1}\left[c_i L_i(\mathbf{x}_{-i})\right] \tag{5.9}$$

Hence we can write:

$$P(H_i \mid \mathbf{X}_{-i} = \mathbf{x}_{-i}) = \frac{a_i}{c_i} \int_{H_{e,i}} \varphi(u_i) \, du_i \tag{5.10}$$

where a_i is a proportionality constant and $\varphi(.)$ is the standard normal PDF. It follows that:

$$\begin{aligned} P(H_i) &= \frac{a_i}{c_i} \int_{\mathbf{x}_{-i}} P(H_i \mid \mathbf{X}_{-i} = \mathbf{x}_{-i}) f_{\mathbf{X}_{-i}}(\mathbf{x}_{-i}) \, d\mathbf{x}_{-i} \\ &= \frac{a_i}{c_i} \int_{H_{e,i}} f_{\mathbf{X}_+}(\mathbf{x}_+) \, d\mathbf{x}_+ \end{aligned} \tag{5.11}$$

where $\mathbf{X}_+ = [\mathbf{X}_{-i}, U_i]^T$ and $f_{\mathbf{X}_+}(.)$ is the joint PDF of the random variable space $\mathbf{X}_+$. Similarly, we can derive that

$$P(F \cap H_i) = \frac{a_i}{c_i} \int_{F \cap H_{e,i}} f_{\mathbf{X}_+}(\mathbf{x}_+) d\mathbf{x}_+ \quad (5.12)$$

The conditional probability of F given H_i is thus

$$P(F|H_i) = \frac{P(F \cap H_i)}{P(H_i)} = \frac{\int_{F \cap H_{e,i}} f_{\mathbf{X}_+}(\mathbf{x}_+) d\mathbf{x}_+}{\int_{H_{e,i}} f_{\mathbf{X}_+}(\mathbf{x}_+) d\mathbf{x}_+} \quad (5.13)$$

where it is noted that the proportionality constant a_i vanishes.

Denoting now by $\mathbf{X}_+$ the random vector consisting of all auxiliary standard normal random variables U_i, corresponding to all information events H_i, and the variables $\mathbf{X}_{-m}$ that are not eliminated when formulating the likelihood functions according to Eq. (5.7), we can express the probability of failure conditional on all information events H_i, as follows:

$$P(F|H_1 \cap ... \cap H_m) = \frac{P(F \cap H_1 \cap ... \cap H_m)}{P(H_1 \cap ... \cap H_m)}$$

$$= \frac{\int_{F \cap H_{e,1} \cap ... \cap H_{e,m}} f_{\mathbf{X}_+}(\mathbf{x}_+) d\mathbf{x}_+}{\int_{H_{e,1} \cap ... \cap H_{e,m}} f_{\mathbf{X}_+}(\mathbf{x}_+) d\mathbf{x}_+} \quad (5.14)$$

As shown in Eqs. (5.13) and (5.14), this approach reduces the reliability updating problem to the evaluation of two probability integrals that can be performed by application of any reliability method, including simulation methods. This is the main advantage of this method, compared to the one discussed in Section 5.2.1, which results in an approximation of the conditional probability.

6 Numerical examples

This chapter presents a number of numerical examples computed by coupling the reliability methods described in Chapters 4 and 5 with finite element models. For the analysis, the reliability software RELY [96] developed as part of this work and integrated into the SOFiSTiK finite element (FE) software package is used. The examples focus on geotechnical applications with uncertain material parameters. Moreover, the spatial variability of the soil is taken into account, using the random field discretization methods described in Section 2.5, and its influence on the reliability results is studied.

6.1 Reliability analysis of a deep circular tunnel

The first example is a deep circular tunnel surrounded by weak rock with uncertain material properties. The reliability is computed for a serviceability limit-state based on a nonlinear FE model. In addition, the influence of the spatial variability of the material properties on the tunnel's reliability is examined. The results shown in this section have been originally presented in [94] and [97].

6.1.1 Finite element solution of tunnel deformation analysis

The tunnel is assumed to be subjected to a hydrostatic pressure and a uniform support pressure (see Figure 6.1). The tunnel's radius is taken as r = 3.3m and the supporting pressure as p_i = 5000kN/m². The hydrostatic pressure is calculated from the stresses resulting due to an assumed 1100m overburden load. The scope of the numerical model is the evaluation of the vertical displacement at the uppermost point of the tunnel's circumference.

The displacement of the tunnel is determined by a numerical evaluation of the ground-support equilibrium point (e.g. see [56]). To this end, an 80 × 80m bounded block of the surrounding soil is modeled using plain strain finite elements. Figure 6.3(b) shows the FE mesh. The material parameters of the soil are taken as independent random variables, described by probability distributions, as shown in Table 6.1. For the specific weight the normal distribution was chosen, while for the remaining parameters the lognormal and beta distributions were utilized due to their advantage in defining lower and upper bounds. For example, the Young's modulus must take positive values and the Poisson's ratio may not be larger than 0.5.

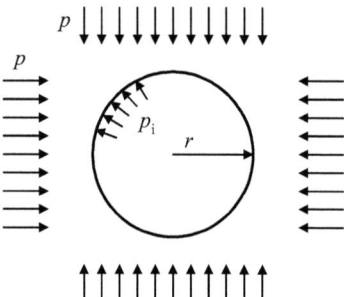

Figure 6.1: Circular tunnel subjected to hydrostatic stress field and uniform support pressure.

The limit-state function is chosen such that the inward displacement of the tunnel's circumference does not exceed a threshold of $u_{x,t}$ = 0.04m:

6.1 Reliability analysis of a deep circular tunnel

$$g(\mathbf{x}) = u_{x,t} - u_{in}(\mathbf{x}) \qquad (6.1)$$

The FE analysis is performed in two steps. First, the modeling of the in-situ hydrostatic stress state is carried out. Then the elements corresponding to the circular tunnel are removed and the supporting pressure is applied by performing a nonlinear computation. The material model used is an elastic-perfectly plastic model with non-associative plastic flow and zero dilatancy, while the yield surface is defined by the Mohr-Coulomb criterion. The choice of the material model is motivated by matching the prerequisites for which an analytical ground reaction curve can be derived [56].

Table 6.1: Material properties.

Parameter	Distribution	Mean	CV
Specific weight γ [kN/m3]	Normal	27.0	5%
Young's modulus E [MPa]	Lognormal	5000.0	25%
Poisson's ratio ν	Beta(0.0, 0.5)	0.2	10%
Friction angle φ [°]	Beta(0.0, 45.0)	35.0	10%
Cohesion c [MPa]	Lognormal	3.7	20%
Dilatancy angle ψ [°]	-	0	-

6.1.2 Results without spatial variability considerations

The reliability analysis was performed applying the FORM with four different optimization algorithms; the HL-RF and its improved version (iHL-RF), as well as the gradient projection (GP) method and the improved GP (iGP), which includes an adaptive adjustment of the step-length utilizing the reduction function of Eq. (4.35). Starting from the mean point, all algorithms were able to converge to a design point, although the reliability index computed by the HL-RF differed slightly from the one computed by the GP algorithms. Table 6.2 shows the reliability index and probability of failure computed and the required number of limit-state function evaluations, for the four methods. It is shown that the standard versions of the algorithms were able to converge faster, while the improved versions required additional function evaluations.

The results were verified by the directional simulation method with 150 deterministic directions (DS-DP) as well as with a number of importance sampling methods, based on the computed design point, namely the standard importance sampling (SIS), the axis orthogonal importance sampling (AOIS) and the directional importance sampling (DIS). For all importance sampling methods, the simulation was carried out with a target coefficient of variation of 5%.

In Table 6.3, the results are compared. The simulation methods produced comparable results with similar performance in terms of efficiency, the AOIS being the most efficient of the bunch. In addition, it is shown that the FORM gives a good approximation, although slightly overestimating the reliability.

Table 6.2: Reliability index and probability of failure computed by FORM.

Optimization method	β	P_f	Number of function evaluations
HL-RF	2.414	7.886×10^{-3}	96
iHL-RF	2.414	7.886×10^{-3}	210
GP	2.410	7.980×10^{-3}	76
iGP	2.410	7.980×10^{-3}	124

Table 6.3: Reliability index and probability of failure computed with various simulation methods.

Method	β	P_f	Number of function evaluations
FORM-GP	2.410	7.980×10^{-3}	76
DS-DP	2.268	1.167×10^{-2}	711
SIS	2.297	1.081×10^{-2}	616
AOIS	2.281	1.128×10^{-2}	522
DIS	2.291	1.098×10^{-2}	726

6.1 Reliability analysis of a deep circular tunnel   157

In Figure 6.2, the influence coefficients, computed as a byproduct of the FORM analysis, are plotted. As discussed in Section 4.3.3, the influence coefficients provide information about the relative importance of the basic random variables, since the random variables are assumed to be independent. As expected, the Young's modulus is the dominant variable, in the sense that it has the biggest influence to the variance of the linearized limit-state function and hence to the reliability of the tunnel. Furthermore, it appears that the specific weight and the friction angle have considerable influence, while the influence of the rest of the variables is negligible. The sign of the coefficients shows whether the corresponding variable is of capacity or demand type. For instance, the influence coefficient corresponding to the Young's modulus E is positive, which means that a positive change of E signifies a positive change of the reliability.

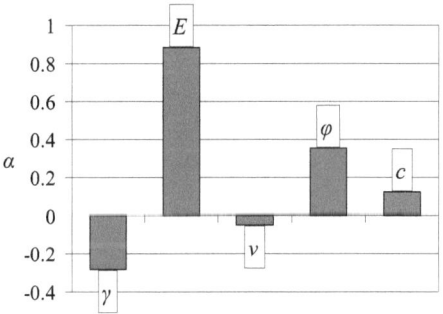

Figure 6.2: Influence coefficients.

6.1.3 Finite element model including spatial variability

It is well known that the material properties of the soil tend to vary in space, even within homogenous layers, which requires the consideration of the soil body as a random field for their proper modeling [105]. Hence, the representation of the material properties by random variables is an approximation, since it implies a perfectly correlated random field. We now examine the influence of the spatial variability of the most important random variables, i.e. the variables with the largest influence coefficients, on the tunnel's reliability. To this end, we model the Young's modulus and friction angle by homogeneous non-Gaussian random fields. To model the joint distribution of the random fields, the Nataf distribution is applied, with marginal distributions as shown in Table 6.1. Moreover, the following isotropic exponential auto-correlation coefficient function is used:

$$\rho_{xx}(\tau) = \exp\left(-\frac{\tau}{l}\right) \qquad (6.2)$$

where τ is the Euclidean distance between two locations and l the correlation length.

The random fields are discretized by the midpoint method using a stochastic finite element (SFE) mesh, consisting of 113 deterministic finite element patches. The selection of the SFE mesh is based on the expected rate of fluctuation which in the case of homogeneous fields is described by the auto-correlation coefficient function (see Section 2.5.4). However, the SFE mesh should be coarse enough to avoid near perfect correlation between the elements, which may cause instability in the probabilistic transformation [24]. Obviously, these requirements differ considerably from the ones for the selection of the deterministic FE mesh. Here, the SFE mesh is chosen first and each SFE is defined in the SOFiSTiK program as a structural area with random material properties. Then, the structural areas are meshed by the SOFiSTiK program, resulting in the deterministic FE mesh. In Figure 6.3, the stochastic and deterministic FE meshes are shown. Figure 6.4 depicts plots of realizations of the lognormal random field representing the Young's modulus, corresponding to the same realization at the equivalent standard normal space, for two different correlation lengths.

6.1 Reliability analysis of a deep circular tunnel   159

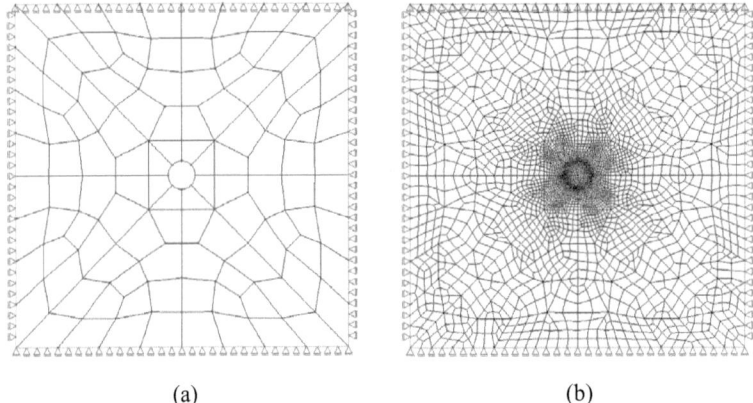

(a) (b)
Figure 6.3: (a) Stochastic FE mesh. (b) Deterministic FE mesh.

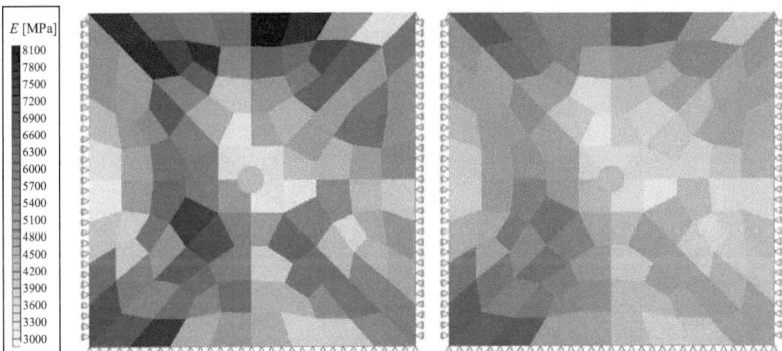

Figure 6.4:   Realizations of the homogeneous lognormal random field representing the Young's modulus E of the soil (the same realization at the equivalent standard normal space) for two different correlation lengths. Left: $l = 10$m. Right: $l = 40$m.

6.1.4 Results accounting for spatial variability of soil

The discretization of the considered random fields representing the spatial variability of the Young's modulus E and friction angle φ leads to a total of

226 random variables. In order to deal with the computational cost in the reliability evaluation due to the large number of random dimensions, the subset simulation method, discussed in Section 4.8.1, is applied. The parameters of the algorithm, i.e. the target value of the conditional probabilities p_0 and the number of samples at each subset N, are chosen as $p_0 = 0.1$ and $N = 500$, following the initial suggestions given in [6]. Table 6.4 shows the computed probability of failure and corresponding reliability index for a progressive consideration of the spatial variability of the two material properties for a correlation length of $l = 5$m. It can be observed that the consideration of E, the parameter with the highest influence, improves the reliability estimation significantly, while the improvement from an additional consideration of the less influential φ is much less important. For all cases that account for the spatial variability of at least one material parameter, the computed failure probability is in the order of 10^{-3}. Therefore, three levels were adequate for the estimation of the failure probability with subset simulation, requiring a total of 1400 limit-state function evaluations (see Section 4.8.1).

Table 6.4: Progressive consideration of the spatial variability of φ and E for a correlation length of 5 m.

Random field consideration	β	P_f
All parameters as random variables	2.297	1.108×10^{-2}
Only φ as random field	2.390	8.420×10^{-3}
Only E as random field	2.949	1.600×10^{-3}
E and φ as random fields	3.046	1.160×10^{-3}

In Figure 6.5, the reliability index is plotted for different values of the correlation length l, namely 5, 10, 20, 40 and 80m, and the consideration of the Young's modulus E as a random field. The plot shows that as l increases and thus the random field becomes more correlated, the reliability index approaches the value computed in the fully correlated case, where all parameters are regarded as random variables. Hence, it may be concluded that neglecting the spatial variation of the uncertain material parameters will lead to an underestimation of the tunnel's reliability. This may be quantified in the extreme case of a poorly correlated random field ($l = 5$m) with a rough 0.65 absolute error, corresponding to one order of magnitude in the probability of failure. Such an underestimation of the reliability in the design phase may have considerable economical impact on the structural design.

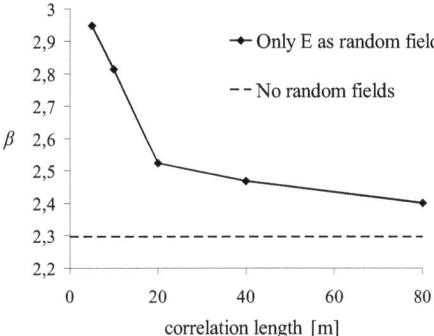

Figure 6.5: Influence of the correlation length of the Young's modulus on the reliability.

6.2 Reliability updating of a cantilever embedded wall

This example is an excavation in sand with a sheet pile retaining wall, where deformation measurements made at an intermediate excavation depth are utilized to update the reliability of the construction site at the stage of full excavation. Therein, uncertainty in the soil material properties is modeled by non-Gaussian random fields. Moreover, the method discussed in Section 5.2.2 is applied to reformulate the measurement information (being of equality-type) into inequality-type information. The structural reliability evaluations required for the Bayesian updating are carried out by means of the subset simulation, in order to cope with the large number of random variables derived from the discretization of the random fields. This example can be found in [98] and [99].

6.2.1 Finite element model

The site consists of a 5.0m deep trench with cantilever sheet piles, without anchors or bottom support (Figure 6.6), in a homogeneous soil layer of dense cohesionless sand with uncertain spatially varying mechanical properties. The soil is modeled in 2D with plane-strain finite elements. For simplicity, neither groundwater nor external loading is considered. Additionally, we take advantage of the symmetry of the trench and model just one half of the soil profile. However, it should be noted that this is an approximation when randomness in the soil material is taken into account. The material model used is an elasto-plastic model with a prismatic yield surface according to the Mohr-Coulomb criterion and a non-associated plastic flow. The probabilistic models of the material properties of the soil are shown in Table 6.5. The spatial variability of the soil is modeled by homogeneous random fields, with the following exponential auto-correlation coefficient function:

$$\rho_{XX}(\mathbf{\tau}) = \exp\left(-\frac{\tau_x}{l_x} - \frac{\tau_z}{l_z}\right) \qquad (6.3)$$

where $\mathbf{\tau} = [\tau_x, \tau_z]^T$ is the vector of absolute distances in the x (horizontal) and z (vertical) directions. The correlation lengths are $l_x = 20$m and $l_z = 5$m for all uncertain soil material properties: specific weight γ, Young's modulus E and friction angle φ. An infinite correlation length is intrinsically assumed in the y direction (out of plane). Cross-correlation between the different material properties is not included. The joint distribution of the random variables in the random fields is the Nataf distribution, with marginal distributions according to Table 6.5. The random fields are discretized by the midpoint method using a SFE mesh, consisting of 144 deterministic FE patches. The stochastic discretization resulted in a total of 3 x 144 = 432 basic random variables. In Figure 6.7, the stochastic and deterministic FE meshes are shown. In Figure 6.8, two realizations of the lognormal random field representing the uncertainty of the Young's modulus are shown.

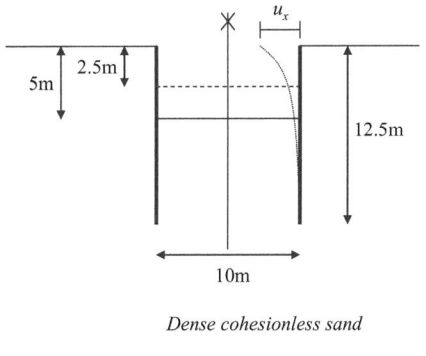

Figure 6.6: Sheet pile wall in sand.

Table 6.5: Material properties.

Parameter	Distribution	Mean	CV
Specific weight γ [kN/m^3]	Normal	19.0	5%
Young's modulus E [MPa]	Lognormal	125.0	25%
Poisson's ratio ν	-	0.35	-
Friction angle φ [°]	Beta(0.0, 45.0)	35.0	10%
Cohesion c [MPa]	-	0.0	-
Dilatancy angle ψ [°]	-	5.0	-

The sheet pile dimension and profile is determined analytically using the conventional method for cantilever sheet pile design in granular soils, which requires equilibrium of the active and passive lateral pressures [135]. Applying a global safety factor of 1.5, the design results in sheet piles of depth of 7.5m and profile PZC 13. The Young's modulus of steel is taken as 210 GPa. The pile is modelled using beam elements with an equivalent rectangular cross section that behaves equally to the sheet pile in bending and axial resistance. The interaction between the retaining structure and the surrounding soil is modelled using nonlinear interface elements. An

164   6 Numerical examples

elastoplastic model with a yield surface defined by the Mohr-Coulomb criterion is used to describe the interface behavior. The elastic properties of the interface elements are taken from the mean values of the adjacent soil, while the strength properties are reduced by the factor 2/3 and a zero dilatancy is chosen.

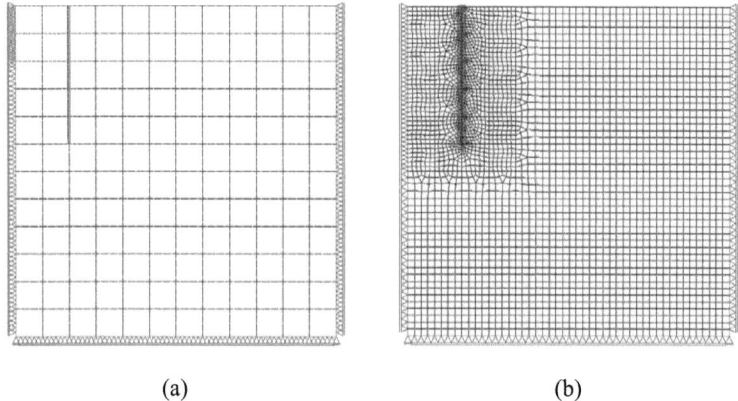

(a) (b)
Figure 6.7: (a) Stochastic FE mesh. (b) Deterministic FE mesh.

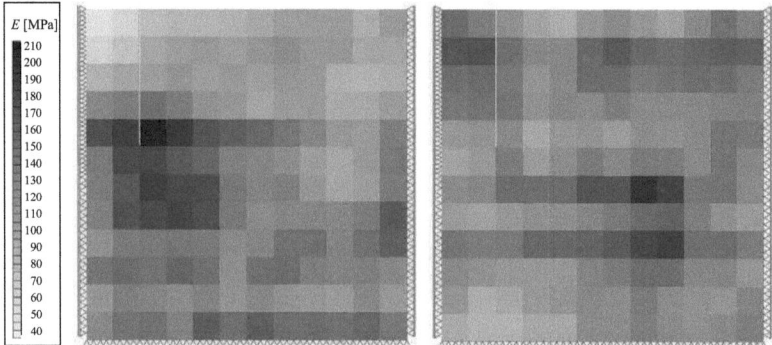

Figure 6.8:   Two realizations of a homogeneous anisotropic lognormal random field representing the Young's modulus E of the soil.

The finite element analysis is performed stepwise, following the construction process. First, the modelling of the in-situ stress state is carried out by means of the K_0-procedure, where K_0 is the lateral earth pressure coefficient at rest, computed here using the expression proposed by Jaky [63] for normally consolidated soils:

$$K_0 = 1 - \sin\varphi \tag{6.4}$$

Next, the sheet pile is installed by activating the corresponding beam and interface elements. Finally, the excavation is modelled by removing the plane-strain elements corresponding to the trench and applying the necessary loading to establish equilibrium.

6.2.2 Limit-state functions

The maximum horizontal u_x displacement occurs at the top of the trench. The failure event F is defined as the event of u_x exceeding a threshold of $u_{x,t} = 0.1$m. Mathematically, this is expressed through the following limit state function:

$$g(\mathbf{x}) = u_{x,t} - u_x(\mathbf{x}) \tag{6.5}$$

This is a serviceability limit state, reflecting the assumed serviceability design requirements. A stability analysis, performed by application of the shear strength reduction technique ([81], see also Section 6.3.1) with the mean values of the random fields, resulted in a factor of safety of 2.5. In Eq. (6.5), $u_x(\mathbf{x})$ is evaluated by the FE analysis for given values of the random variables $\mathbf{X}$. We assume that a measurement of the displacement u_x is made at an intermediate excavation step of 2.5m depth. This information is expressed by an event H, described by the following likelihood function:

$$L(\mathbf{x}) = \varphi\left\{\left[u_{x,m} - u_x(\mathbf{x})\right]/\sigma_{\varepsilon,m}\right\}/\sigma_{\varepsilon,m} \tag{6.6}$$

where $\sigma_{\varepsilon,m}$ is the standard deviation of the measurement error, which is a zero mean Gaussian random variable; $\varphi(.)$ is the standard normal PDF. The

166   6 Numerical examples

corresponding equivalent inequality limit state function is obtained according to Eq. (6.7):

$$h_e(\mathbf{x},u) = u_a - \Phi^{-1}\left[\frac{c}{\sigma_{\varepsilon,m}}\varphi\left(\frac{u_{x,m} - u_x(\mathbf{x})}{\sigma_{\varepsilon,m}}\right)\right] \quad (6.7)$$

where u_a is the realization of the auxiliary standard normal random variable. The constant is chosen as $c = \sigma_{\varepsilon,m}$, which satisfies the condition $cL(\mathbf{x}) \leq 1$ (refer to Section 5.2.2).

6.2.3 Results

Figure 6.9 depicts the deformed configuration at the final excavation stage computed with the mean values of the random fields. This analysis gives a first-order approximation of the mean (expected) displacements. At the top of the trench, the estimated mean value of the horizontal displacement u_x is 50.2mm.

The reliability analysis is performed by means of subset simulation with parameters $p_0 = 0.1$ and $N = 500$. Without measurements, the computed failure probability is $P(F) = 1.36 \times 10^{-2}$ with a corresponding reliability index $\beta = 2.21$.

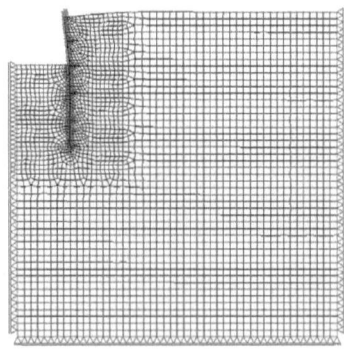

Figure 6.9: Magnified deformed configuration at the final excavation step.

6.2 Reliability updating of a cantilever embedded wall

For the estimation of the updated failure probability conditional on the measurement event H, the integrals in Eq. (5.13) were evaluated with subset simulation. The reliability updating was performed for different measurement outcomes $u_{x,m}$, and different values of the standard deviation $\sigma_{\varepsilon,m}$ of the measurement error. The results are summarized in Table 6.6 and the computed reliability indices are plotted in Figure 6.10. For comparison, the (a-priori) first-order approximation of the expected value of the measurement outcome $u_{x,m}$ is computed as 2.6mm.

Not surprisingly, for measurements significantly higher than the expected value, the updated failure probability is higher than the prior probability. This difference is more pronounced when the measurement device is more accurate, i.e. when $\sigma_{\varepsilon,m}$ is smaller. For measurements lower than the expected value, the updated failure probability is lower than the prior probability. Again, the difference increases with decreasing value of $\sigma_{\varepsilon,m}$, because this implies a higher information content of the measurement. It is noted that a measurement that corresponds exactly to the expected value of the deformation would lead to a posterior failure probability that is lower than the prior probability, due to a reduction of uncertainty.

The analysis assumes in-plane symmetry. This is a valid assumption in the absence of spatial variability considerations. However, for the present example this assumption neglects asymmetric realizations of the random fields, which can possibly influence the computed reliability. Moreover, since the analysis is performed in 2D, the correlation length in the out-of-plane direction is assumed to be infinite.

The number of deterministic FE analyses required by the subset simulation ranges between 1900 and 3700, which includes the evaluation of both integrals in Eq. (5.13). The higher amount of computations is observed in the case where the assumed measurement differs considerably from the expected value (i.e. the case where $u_{x,m}$ = 10mm). This is due to the small value of the probability $P(H_e)$ in Eq. (5.13), resulting in a larger number of levels M in the corresponding run of the subset simulation algorithm.

168   6 Numerical examples

Table 6.6:   Updated failure probability and reliability index.

Measurement	$\sigma_{\varepsilon,m} = 2$mm		$\sigma_{\varepsilon,m} = 1$mm	
	$P(F\|H)$	β	$P(F\|H)$	β
$u_{x,m} = 10$ mm	2.18×10^{-1}	0.78	3.31×10^{-1}	0.44
$u_{x,m} = 5$ mm	2.09×10^{-2}	2.04	3.59×10^{-2}	1.80
$u_{x,m} = 2$ mm	6.74×10^{-3}	2.47	1.84×10^{-3}	2.90

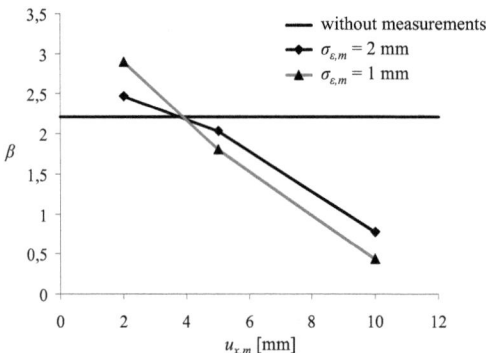

Figure 6.10: Reliability index against measured displacement.

For practical implementation, the reliability can be computed conditional on different hypothetical measurement outcomes, prior to the in-situ measurement. Then a threshold value for the actual measurement may be obtained as a function of the target reliability index β_t as illustrated in Figure 6.11. Assuming that the target reliability is $\beta_t = 2.5$ and the measurement accuracy is $\sigma_{\varepsilon,m} = 1$mm, the threshold value is 3.1mm. Any measurement larger than this value corresponds to a reliability index less than the acceptable one. This would indicate that the retaining wall would not satisfy the reliability requirements at the final excavation stage and additional measures (e.g. anchors) would be necessary.

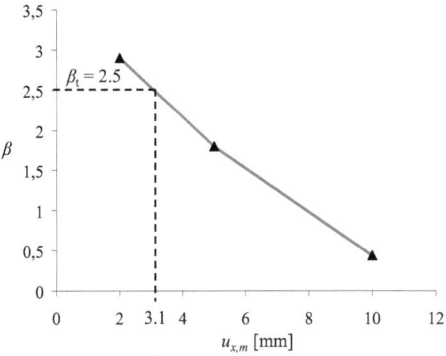

Figure 6.11: Evaluation of a threshold measurement value for an accepted reliability index $\beta_t = 2.5$ and a measurement device with $\sigma_{\varepsilon,m} = 1$mm.

6.3 Reliability-based design of slope angle

In the practical situation of a trench excavation, the geotechnical engineer may decide upon the desired target slope reliability, depending on the particular safety requirements. This example describes a procedure for the design of the slope angle given the target reliability index. To this end, the inverse FORM discussed in Section 4.4 is applied along with the shear strength reduction method for finite element slope stability analysis [81]. The spatial variability of the soil is included by embedding the slope profile in a standard domain and performing a series expansion of the relevant random fields in that domain, i.e. applying the embedded domain discretization approach presented in Section 2.5.5. This approach allows the definition of a consistent representation of the spatial variability, independent of the changes of the geometry of the slope profile as the slope angle is varied in the process of finding the optimal solution. A selection of the results presented in the section can be found in [93], while a similar application is presented in [131].

6.3.1 Finite element slope stability analysis

The example consists of a homogeneous slope with a foundation layer ($H = $ 5m, $D = $ 10m, see Figure 6.12). The slope angle θ can be selected by the designer. Clearly, an increase in the slope angle will lead to a reduction of the cost for the excavation of the trench, but also to a decrease of slope stability. The slope is modeled in 2D with plain strain elasto-plastic elements, with a yield surface governed by the Mohr-Coulomb failure criterion. The FE mesh is shown in Figure 6.13. The elasto-plastic deformations are computed as the converged pseudo time-dependent elasto-viscoplastic solution, applying the viscoplastic strain method ([124], [141]).

The material properties of the soil are given in Table 6.7. The considered random variables are those relevant to shear failure, i.e. the friction angle φ and the cohesion c. Both variables are modeled by lognormal random fields, say $X_i(\mathbf{t})$, $i = \varphi, c$, where $\mathbf{t}$ stands for the location vector. Utilizing the Nataf distribution, we can define a marginal transformation of the lognormal field $X_i(\mathbf{t})$ to an equivalent Gaussian field $U_i(\mathbf{t})$, as follows (see Section 2.4.6):

$$U_i(\mathbf{t}) = \ln(X_i(\mathbf{t})) \qquad (6.8)$$

In the case where data are available, the correlation structure of the underlying Gaussian fields $U_i(\mathbf{t})$ can be estimated. Here, it is assumed that the fields $U_i(\mathbf{t})$ have the exponential autocorrelation coefficient function of Eq. (6.3) and the correlation length is considered identical in both directions.

The fields $U_i(\mathbf{t})$ are discretized by the Karhunen-Loève (KL) expansion, applying the embedded-domain approach, presented in Section 2.5.5. The approach utilizes the rectangular domain Ω_r, shown in Figure 6.14, for the solution of the Fredholm eigenvalue problem. Moreover, the selected form of the auto-correlation coefficient function allows a closed form solution to the eigenvalue problem in the rectangular domain Ω_r [41].

6.3 Reliability-based design of slope angle 171

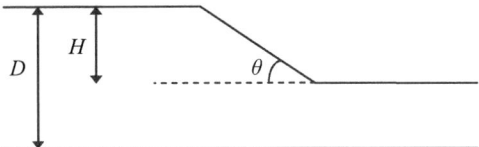

Figure 6.12: Slope profile.

Table 6.7: Material properties.

Parameter	Distribution	Mean	CV
Specific weight γ [kN/m^3]	-	20.0	-
Young's modulus E [MPa]	-	100.0	-
Poisson's ratio v	-	0.3	-
Friction angle φ [°]	Lognormal	12	10%
Cohesion c [kPa]	Lognormal	15.0	20%
Dilatancy angle ψ [°]	-	0	-

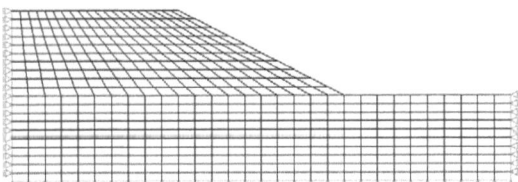

Figure 6.13: Finite element mesh of the example slope.

Note that in the FE modeling process, we take advantage of the symmetry of the trench and therefore we model just one half of the soil profile, as shown in Figure 6.13. However, this consideration becomes an approximation when randomness in the soil material is taken into account.

For the sake of simplicity, a deterministic value of $\psi = 0$ is taken, corresponding to a non-associated flow rule. It has been shown that this selection for the value of the dilatancy angle yields reliable safety factors [49]. For the remaining material parameters (E, v and γ), the deterministic values listed in Table 6.7 are considered, since the influences of their uncertainties on the stability of the slope are insignificant.

The factor of safety of the slope is computed applying the shear strength reduction technique [81]. According to this approach, the factor of safety (FS) is defined as the number by which the original strength parameters (the tangent for the friction angle) must be divided in order to bring the slope to the failure state. This definition is strictly equivalent to the classical definition of the factor of safety used in the limit-equilibrium methods (e.g. see [37]). Denoting the factored strength parameters by φ_f, c_f, we have:

$$\varphi_f = \arctan\left(\frac{\tan(\varphi)}{FS}\right) \qquad (6.9)$$

$$c_f = \frac{c}{FS} \qquad (6.10)$$

For a given realization of the soil properties, the FS is computed by applying a stepwise procedure, whereby the strength parameters are gradually reduced by an increasing factor FS_i and an elasto-plastic FE computation is performed at each step i. The procedure continues until the factor FS_i reaches a value at which failure occurs. Failure is defined in this study as the failure of convergence of the viscoplastic strain algorithm after a given maximum number of iterations.

The limit-state function, with negative values defining the failure condition of the slope, is given by:

$$g(\mathbf{x},\theta) = FS(\mathbf{x},\theta) - 1 \qquad (6.11)$$

Starting from an initial slope $\theta = \arctan(0.5) \approx 26.6°$ (corresponding to a factor of safety FS = 1.77 for the mean properties of the soil), we aim at finding the optimal slope angle for a target reliability index of $\beta_t = 3.4$. To this end, we apply the inverse FORM procedure described in Section 4.4.

It should be noted that the description of the random fields $U_i(\mathbf{z})$ is not affected by the change of the geometry that takes place due to the change of the slope angle θ at each iteration step of the inverse FORM algorithm.

This is due to the fact that the discretization of $U_i(\mathbf{z})$ is based on the rectangular domain Ω_r and therefore is independent of the actual domain Ω (see Figure 6.14).

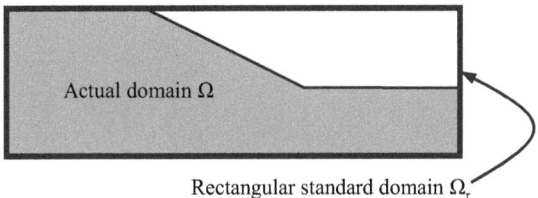

Figure 6.14: Embedded domain discretization of the random fields describing the strength parameters.

6.3.2 Results

In a first step, an infinite correlation length is selected, so that the random fields reduce to random variables. The inverse FORM algorithm converged in 5 iteration steps with the solution $\theta = 34.8°$, which corresponds to a FS = 1.58 for the mean soil material properties. Figure 6.15 shows the deformed mesh at failure (i.e. with the mean values of the strength parameters reduced until failure occurs) for the initial and design values of θ.

Spatial variabilities of the friction angle and cohesion are taken into account for three different correlation lengths: $l = 5$m, 10m and 20m. To validate the applicability of the embedded-domain (KL-ED) approach, we compare its convergence to the one of the standard FE approach for solving the Fredholm eigenvalue problem in the actual domain (KL-FE) (see Section 2.5.3), using the FE mesh shown in Figure 6.13. The error measure $err(\mathbf{t})$ used in the convergence study is the variance of the truncation error of the KL expansion divided by the variance of the original field [see Eq. (2.163)]. In Figure 6.16, the spatial average of $err(\mathbf{t})$ over the domain Ω is plotted against the number of terms in the KL expansion, for both methods. As expected, the smaller error at any order of expansion is obtained by the standard KL-FE approach, since the approximated eigenfunctions are optimal for that domain. However, the errors obtained by the two approaches at each order of expansion do not differ significantly.

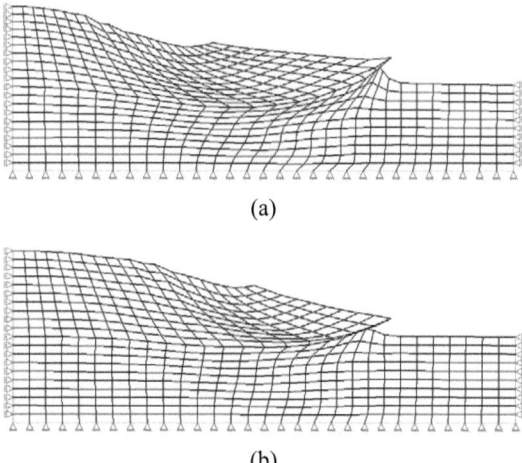

Figure 6.15: Deformed mesh at failure for (a) the initial design $\theta = 26.6°$ and (b) the final design when the spatial variability is neglected $\theta = 34.8°$.

The number of random variables used for the ED representation of each random field is 3, 6 and 18 for the cases with $l = $ 20m, 10m and 5m, respectively. This selection is based on the requirement that the spatial average of the relative variance of the truncation error of the field be less than 0.3. This requirement is verified in Table 6.8, where the computed reliability index β for the design slope of Figure 6.15 is compared to the number of terms in the KL-ED expansion for the case where $l = 10$m. Table 6.8 shows that 6 terms in the expansion yield a reasonable approximation of β.

6.3 Reliability-based design of slope angle

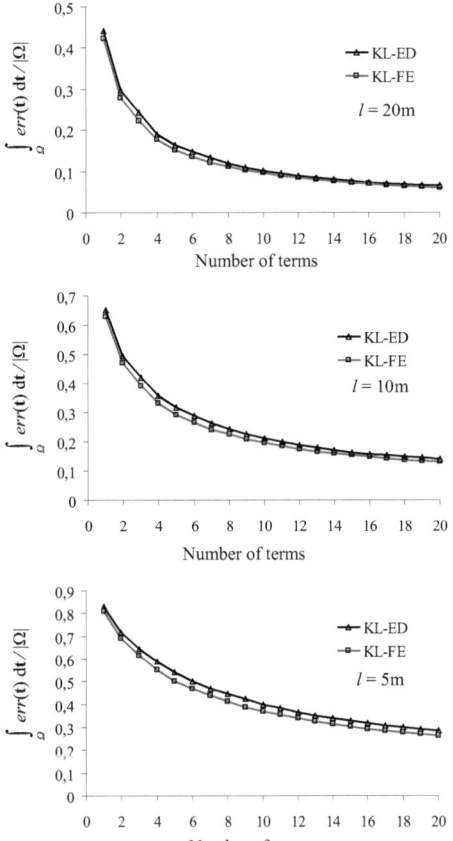

Figure 6.16: Spatial average of error variance against number of terms in the expansions.

For each correlation length, we consider three different cases: in the first two, only the spatial variability of one of the two soil strength parameters is taken into account, while the other is modeled as a random variable; in the third case the spatial variability of both parameters is considered with equal correlation lengths. The inverse FORM converged in all cases within 5-7 iteration steps. In Figure 6.17, the optimal slope angle is plotted against the correlation length for a progressive consideration of the spatial variability of c and φ. Figure 6.18 shows the corresponding factors of safety computed for the mean values of the soil properties at each design. The results indicate that the cohesion c is the most influential parameter, in this particular slope. Moreover, the consideration of the spatial variability of cohesion leads to significantly larger values of the optimal slope angle (respectively smaller values of the required factor of safety) compared to the one obtained, when the soil properties are considered to be random variables. As an example, in the case $l = 5$m, neglecting the spatial variability of the cohesion leads to a 20% smaller reliability-based slope design angle.

Table 6.8: Influence of the number of terms in the KL-ED expansion on the reliability index β for $l = 10$ m.

Order of KL-ED expansion	Spatial average of error variance	β
1	0.650	5.285
2	0.496	4.237
3	0.419	4.095
4	0.357	3.999
5	0.316	3.919
6	0.288	3.854
7	0.263	3.843
8	0.242	3.835
9	0.225	3.832
10	0.211	3.831

In Figure 6.19, the realization of the cohesion at the most probable failure point is plotted for the final design and for the three different considered correlation lengths. The plots show that low values of the cohesion are concentrated at the location where the failure is initiated. In addition, failure initiation becomes more local as the correlation length becomes smaller. Conversely, for large correlation lengths, weak material is distributed throughout the domain at failure. This implies that a small correlation length allows local loss of strength to lead to failure, since the event of weak material in a specific location of an otherwise high strength soil is part of the sample space. However, a large number of such events that do not lead to failure will also be part of the sample space. On the other hand, the possibility of events that can lead to localized failure decreases as the correlation length becomes larger. This explains the finding that for small correlation lengths more reliable slopes and, therefore, less conservative slope angle designs arise.

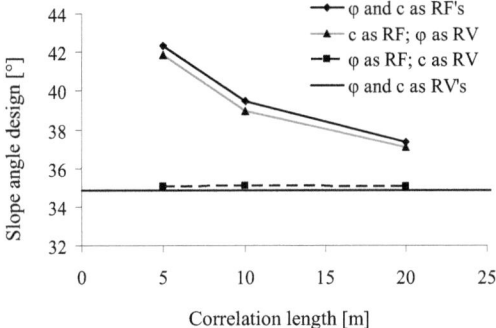

Figure 6.17: Influence of the spatial variability of φ and c on the reliability-based slope angle design.

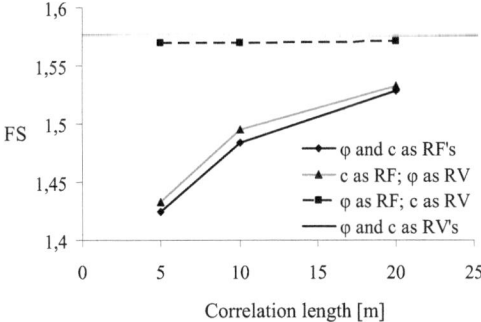

Figure 6.18: Influence of the spatial variability of φ and c on the factor of safety FS at the reliability-based design.

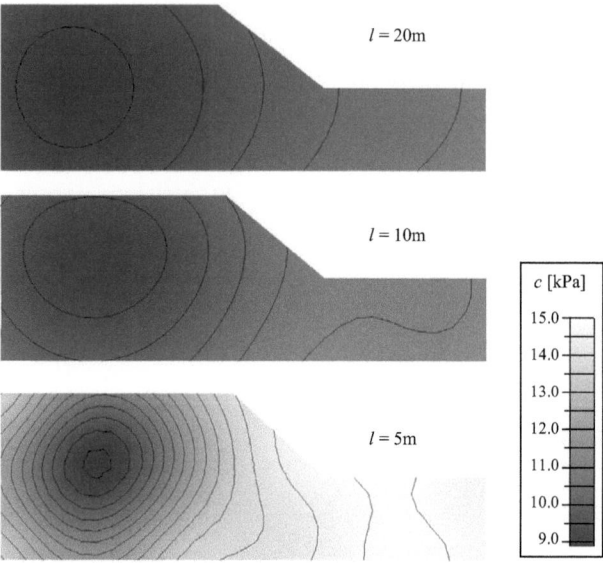

Figure 6.19: Realization of the cohesion at the most probable failure point for the final design.

7 Conclusion

This book presented mathematical and numerical concepts for the modeling of uncertain quantities in structural systems and provided a detailed review of methods for non-intrusive finite element (FE) structural reliability assessment. These methods have been implemented in a tool that is coupled with a commercial FE software package. A summary of the main outcomes of this work throughout its chapters is given bellow.

- Chapter 2 collected the basic concepts of the theory of probability and random variables that are needed for the modeling of uncertainties. The second part of the chapter focused on the representation of the spatial variability of uncertain quantities by application of random field theory. An overview of the existing random field discretization methods, i.e. methods for the representation of the random field using a finite number of random variables, was presented. The methods were compared in terms of their efficiency and advantages as well as drawbacks of each method were specified. Moreover, a novel discretization approach for the treatment of random fields defined on non-standard domains was proposed. The method is based on the Karhunen-Loève expansion, but solves the associated integral eigenvalue problem in an embedded rectangular volume. The validity of this approach was verified by its convergent behavior for several numerical examples. The method can be advantageous in FE models with complex domains, where the numerical solution of the eigenvalue problem involves considerable computational cost, or for domains that change throughout the computation.

- Chapter 3 presented the basics on reliability analysis. Therein, the definitions of the limit-state function and the generalized reliability index were given. Moreover, the system reliability problem and the corresponding equivalent definition of the limit-state function were introduced.

- Chapter 4 presented a review of the implemented FE reliability analysis methods. The first part of the chapter focused on approximation methods (FORM/SORM) and their application to both component and system reliability problems. The determination of the most probable failure point (also known as design point), required for the FORM/SORM, involves the solution of an equality-constrained quadratic optimization problem. Several existing optimization algorithms were presented and a novel approach, based on an adaptive selection of the step-length for the gradient projection method was introduced. The method presented a robust convergence behavior for the examples considered, overcoming the well known instability of the standard gradient projection algorithm. The second part of the chapter provided a detailed presentation of existing simulation methods with a focus on advanced sampling techniques for the reduction of the variance of the probability estimate. A conditional importance sampling method was introduced that improved the performance of the standard importance sampling. Moreover, an adaptive directional importance sampling method was proposed based on an initial simulation with a set of uniformly distributed deterministic directions. This approach has the advantage that the evaluation of the design point is not needed and thus it can be used in cases where the FORM

6.3 Reliability-based design of slope angle

optimization algorithm fails to converge. Next, simulation of problems involving a large number of random variables was treated and an existing approach, namely the subset simulation, was presented in detail. The final part of the chapter presented the implemented response surface methods that create a surrogate model of the FE solver, based on an experimental design.

- Chapter 5 introduced the problem of updating the reliability estimate conditional on available information. Special care must be taken for problems where the information is of equality-type, e.g. measurement of a system characteristic. Two existing methods that are able to approach such problems were reviewed.

- Chapter 6 presented three industrial applications that demonstrated the applicability of the methods described in the previous chapters. The first example examined the performance of various reliability methods for the reliability analysis of a deep circular tunnel surrounded by weak rock. The second example proposed an approach for the reliability updating of geotechnical sites conditional on measurement information obtained in-situ. The proposed approach combined random field modeling of the spatial variability of the soil material, a recently proposed reliability updating method that is able to handle measurement information and the subset simulation method for coping with the large number of random variables arising from the discretization of the random fields. The third example presented a methodology for the reliability-based design of the slope angle accounting for the spatial variability of the soil. This approach combined an inverse FORM algorithm with the embedded-domain method for the discretization of the random fields. The latter method allowed a constant discretization throughout the inverse FORM optimization procedure, since the random field description is not affected by the change of the geometry of the slope profile that takes place at each iteration step.

Bibliography

[1] Ambartzumian, R., Der Kiureghian, A., Ohanian, V., Sukiasian, H. Multinormal probability by sequential conditioned importance sampling: theory and application. *Probabilistic Engineering Mechanics* 13(4) (1998), pp. 299-308.

[2] Ang, A. H.-S., Tang, W. H. *Probability Concepts in Engineering: Emphasis on Applications to Civil and Environmental Engineering.* 2nd edition. Hoboken, NJ: John Wiley & Sons, 2007.

[3] Ang, G. L., Ang, A. H.-S., Tang, W. H. Optimal importance sampling density estimator. *Journal of Engineering Mechanics, ASCE* 118(6) (1992), pp. 1146-1163.

[4] Atkinson, K. E. *The Numerical Solution of Integral Equations of the Second Kind.* Cambridge, UK: Cambridge University Press, 1997.

[5] Au, S. K., Beck, J. L. A new adaptive importance sampling scheme for reliability calculations. *Structural Safety* 21 (1999), pp. 135-158.

[6] Au, S. K., Beck, J. L. Estimation of small failure probabilities in high dimensions by subset simulation. *Probabilistic Engineering Mechanics* 16(4) (2001), pp. 263-277.

[7] Bjerager, P. Probability integration by directional simulation. *Journal of Engineering Mechanics, ASCE* 114(8) (1988), pp. 1285-1302.

[8] Bjerager, P., Krenk, S. Parametric sensitivity in first-order reliability theory. *Journal of Engineering Mechanics, ASCE* 115(7) (1989), pp. 1577-1582.

[9] Boggs, P. T., Tolle, J. W. Sequential quadratic programming. *Acta Numerica* 4 (1995), pp. 1-51.

[10] Box, G. E. P., Muller, M. E. A note on the generation of random normal deviates. *Annals of Mathematical Statistics* 29 (1958), pp. 610-611.

[11] Breitung, K. Asymptotic approximations for multinormal integrals. *Journal of Engineering Mechanics, ASCE* 110(3) (1984), pp. 357-366.

[12] Bucher, C. Adaptive sampling - an iterative fast Monte Carlo procedure. *Structural Safety* 5 (1988), pp. 119-126.

[13] Bucher, C. Asymptotic sampling for high-dimensional reliability analysis. *Probabilistic Engineering Mechanics* 24 (2009), pp. 504-510.

[14] Bucher, C., Bourgund, U. A fast and efficient response surface approach for structural reliability problems. *Structural Safety* 7 (1990), pp. 57-66.

[15] Cheng, J., Li, Q. S. Reliability analysis of structures using artificial neural network based genetic algorithms. *Computer Methods in Appied Mechanics and Engineering* 197 (2008), pp. 3742-3750.

[16] Cornell, C. A. A probability-based structural code. *Journal of the American Concrete Institute* 66(12) (1969), pp. 974-985.

[17] Dai, H., Wang, W. Application of low-discrepancy sampling method in structural reliability analysis. *Structural Safety* 31 (2009), pp. 55-64.

[18] Deák, I. Three digit accurate multiple normal probabilities. *Numerische Mathematik* 35 (1980), pp. 369-380.

185   Bibliography

[19] Deng, J., Gu, D., Li, X., Yue, Z. Q. Structural reliablity analysis for implicit performance functions using artificial neural network. *Structural Safety* 27 (2005), pp. 25-48.

[20] Der Kiureghian, A. First- and second- order reliability methods. Chapter 14 in *Engineering Design Reliability Handbook*, Nikolaidis, E., Ghiocel, D. M., Singhal, S. (eds.), Boca Raton, FL: CRC Press, 2005.

[21] Der Kiureghian, A., Dakessian, T. Multiple design points in first and second-order reliability. *Structural Safety* 20 (1998), pp. 37-49.

[22] Der Kiureghian, A., de Stefano, M. Efficient algorithms for second-order reliability analysis. *Journal of Engineering Mechanics, ASCE* 117(12) (1991), pp. 2904-2923.

[23] Der Kiureghian, A., Ditlevsen, O. Aleatory or epistemic? Does it matter? *Structural Safety* 31 (2009), pp. 105-112.

[24] Der Kiureghian, A., Ke, J.-B. The stochastic finite element method in structural reliability. *Probabilistic Engineering Mechanics* 3(2) (1988), pp. 83-91.

[25] Der Kiureghian, A., Liu, P.-L. Structural reliability under incomplete probability information. *Journal of Engineering Mechanics, ASCE* 112(1) (1986), pp. 85-104.

[26] Der Kiureghian, A., Zhang, Y. Space variant finite element reliability analysis. *Computer Methods in Appied Mechanics and Engineering* 168 (1999), pp. 173-183.

[27] Der Kiureghian, A., Zhang, Y., Li, C.-C. Inverse reliablity problem. *Journal of Engineering Mechanics, ASCE* 120(5) (1994), pp. 1154-1159.

[28] Deodatis, G. Bounds on response variability of stochastic finite element systems. *Journal of Engineering Mechanics, ASCE* 116(3) (1990), pp. 565-585.

[29] Deodatis, G. The weighted integral method, I : stochastic stiffness matrix. *Journal of Engineering Mechanics, ASCE* 117(8) (1991), pp. 1865-1877.

[30] Deodatis, G. The weighted integral method, II : response variability and reliability. *Journal of Engineering Mechanics, ASCE* 117(12) (1991), pp. 2906-2923.

[31] Deodatis, G. Non-stationary stochastic vector processes: seismic ground motion applications. *Probabilistic Engineering Mechanics* 11(3) (1996), pp. 149-167.

[32] Deodatis, G., Micaletti, R. C. Simulation of highly skewed non-Gaussian stochastic processes. *Journal of Engineering Mechanics, ASCE* 127 (2001), pp. 1284-1295.

[33] Ditlevsen, O. Generalized second moment reliability index. *Journal of Structural Mechanics* 7 (1979), 435-451.

[34] Ditlevsen, O. Narrow reliability bounds for structural systems. *Journal of Structural Mechanics* 7 (1979), 453-472.

[35] Ditlevsen, O., Madsen, H. O. *Structural Reliability Methods*, Chichester, UK: Wiley, 1996.

[36] Ditlevsen, O., Melchers, R. E., Gluver, H. General multi-dimensional probability integration by directional simulation. *Computers and Structures* 36(2) (1990), pp. 355-368.

[37] Duncan, J. M. State of the art: Limit equilibrium and finite element analysis of slopes. *Journal of Geotechnical Engineering, ASCE* 122(7) (1996), pp. 577-596.

[38] Faravelli, L. Response surface approach for reliability analysis. *Journal of Engineering Mechanics, ASCE* 115(12) (1989), pp. 2763-2781.

Bibliography

[39] Fiessler, B., Neumann, H.-J., Rackwitz, R. Quadratic limit states in structural reliability. *Journal of the Engineering Mechanics Division, ASCE* 105(4) (1979), pp. 661-676.

[40] Ghanem, R. G., Spanos, P. D. Spectral stochastic finite element formulation for reliability analysis. *Journal of Engineering Mechanics, ASCE* 117(10) (1991), pp. 2351-2372.

[41] Ghanem, R. G., Spanos, P. D. *Stochastic Finite Elements - A Spectral Approach*. Berlin: Springer-Verlag, 1991.

[42] Gilks, W. R., Richardson, S., Spiegelhalter, D. J. *Markov Chain Monte Carlo in Practice*. London: Chapman & Hall, 1996.

[43] Gollwitzer, S., Rackwitz, R. An efficient numerical solution to the multinormal integral. *Probabilistic Engineering Mechanics* 3(2) (1988), pp. 98-101.

[44] Grigoriu, M. Crossings of non-Gaussian translation processes. *Journal of Engineering Mechanics, ASCE* 110(4) (1984), pp. 610-620.

[45] Grigoriu, M. On the spectral representation method in simulation. *Probabilistic Engineering Mechanics* 8 (1993), pp. 75-90.

[46] Grigoriu, M. *Stochastic Calculus. Applications in Science and Engineering*. New York: Birkhäuser, 2002.

[47] Grigoriu, M. Evaluation of Karhunen-Loève, spectral, and sampling representations of stochastic processes. *Journal of Engineering Mechanics, ASCE* 132(2) (2006), pp. 179-189.

[48] Griffiths, D. V., Huang, J., Fenton, G. A. Influence of the spatial variability on slope stability using 2-D random fields, *Journal of Geotechnical and Geoenvironmental Engineering., ASCE* 135(10) (2009), pp. 1367-1378.

[49] Griffiths, D. V., Lane, P. A. Slope stability analysis by finite elements, *Géotechnique*, 49(3) (1999), pp. 387-403.

[50] Guan, X. L., Melchers, R. E. Effect of response surface parameter variation on structural reliability estimates. *Structural Safety* 23 (2001), pp. 429-444.

[51] Halton, J. H. On the efficiency of certain quasi-random sequences of points in evaluating multi-dimensional integrals. *Numerische Mathematik* 2 (1960), pp. 196.

[52] Harbitz, A. An efficient sampling method for probability of failure calculation. *Structural Safety* 3 (1986), pp. 109-115.

[53] Hasofer, A. M., Lind, N. C. Exact and invariant second moment code format. *Journal of Engineering Mechanics, ASCE* 100 (1974), pp. 111-121.

[54] Hastings, W. K. Monte Carlo sampling methods using Markov chains and their applications. *Biometrika* 57(1) (1970), pp. 97-109.

[55] Haukaas, T., Der Kiureghian, A. Parameter sensitivity and importance measures in nonlinear finite element reliability analysis. *Journal of Engineering Mechanics, ASCE* 131(10) (2005), pp. 1013-1026.

[56] Hoek, E. Reliability of Hoek-Brown estimates of rock mass properties and their impact on design. *International Journal of Rock Mechanics and Mining Sciences* 35(1) (1998), pp. 63-68.

[57] Hohenbichler, M., Rackwitz, R. Non-normal dependent vectors in structural safety. *Journal of Engineering Mechanics, ASCE* 107 (1981), pp. 1227-1238.

[58] Hohenbichler, M., Rackwitz, R. First-order concepts in system reliability. *Structural Safety* 1(3) (1983), pp. 177-188.

[59] Hohenbichler, M., Rackwitz, R. Sensitivity and importance measures in structural reliability. *Civil Engineering Systems* 3 (1986), pp. 203-210.

Bibliography

[60] Hohenbichler, M., Rackwitz, R. Improvements of second-order reliability estimates by importance sampling. *Journal of Engineering Mechanics, ASCE* 114 (1988), pp. 2195-2199.

[61] Hurtado, J. E. An examination of methods for approximating implicit limit state functions from the viewpoint of statistical learning theory. *Structural Safety* 26 (2004), pp. 271-293.

[62] Hurtado, J. E., Alvarez, D. A. Neural-network-based reliability analysis: a comparative study. *Computer Methods in Appied Mechanics and Engineering* 191 (2001), pp. 113-132.

[63] Jaky, J. The coefficient of earth preassure at rest. In Hungarian. *Journal of the Society Hungarian Architects and Engineers* (1948), pp. 355-358.

[64] Katafygiotis, L. S., Cheung, S. H. Application of spherical subset simulation method and auxiliary domain method on a benchmark reliability study. *Structural Safety* 29 (2007), pp. 194-207.

[65] Katsuki, S., Frangopol, D. M. Hyperspace division method for structural reliability. *Journal of Engineering Mechanics, ASCE* 120(11) (1994), pp. 2405-2427.

[66] Keese, A. *Numerical Solution of Systems with Stochastic Uncertainties - A General Purpose Framework for Stochastic Finite Elements*. Doctoral Thesis. Technische Universität Braunschweig, 2004.

[67] Kitterrød, N.-O., Gottschalk, L. Simulation of normal distributed smooth fields by Karhunen-Loève expansion in combination with kriging. *Stochastic Hydrology and Hydraulics* 11 (1997), pp. 459-482.

[68] Koutsourelakis, P. S., Pradlwarter, H. J., Schuëller, G. I. Reliability of structures in high dimensions, part I: algorithms and applications. *Probabilistic Engineering Mechanics* 19(4) (2004), pp. 409-417.

[69] Kuschel, N., Rackwitz, R. Two basic problems in reliability-based structural optimization. *Mathematical Methods of Operations Research* 46 (1997), pp. 309-333.

[70] Lagaros, N. D., Stefanou, G., Papadrakakis, M. An enhanced hybrid method for the simulation of highly skewed non-Gaussian stochastic fields. *Computer Methods in Appied Mechanics and Engineering* 194 (2005), pp. 4824-4844.

[71] Lemaire, M. *Structural Reliability*. Hoboken, NJ: John Wiley & Sons, 2009.

[72] Li, C.-C., Der Kiureghian, A. Optimal discretization of random fields. *Journal of Engineering Mechanics, ASCE* 119(6) (1993), pp. 1136-1154.

[73] Liu, P.-L., Der Kiureghian, A. Multivariate distribution models with prescribed marginals and covariances. *Probabilistic Engineering Mechanics* 1(2) (1986), pp. 105-112.

[74] Liu, P.-L., Der Kiureghian, A. Optimization algorithms for structural reliability. *Structural Safety* 9 (1991), pp. 161-177.

[75] Liu, W. K., Belytschko, T., Mani, A. Random field finite elements. *International Journal on Numerical Methods in Engineering* 23 (1986), pp. 1831-1845.

[76] Loève, M. *Probability Theory, vols I and II*. New York: Springer-Verlag, 1978.

[77] Lovisolo, L., da Silva, E. A. B. Uniform distribution of points on a hypersphere with applications to vector bit-plane encoding. *IEE Proceedings - Vision, Image and Signal Processing* 148 (2001), pp. 187-193.

[78] Luenberger, D. G. *Linear and Nonlinear Programming*. 2nd edition. New York: Springer-Verlag, 2003.

Bibliography

[79] Madsen, H. O. Model updating in reliability theory. In: *Proc. 5th International Conference on Applications of Statistics and Probability in Civil Engineering ICASP 5*, Vancouver, 1987. Waterloo, Ontario: University of Waterloo Press.

[80] Madsen, H. O. Omission sensitivity factors. *Structural Safety* 5 (1988), pp. 35-45.

[81] Matsui, T., San K. C. Finite element slope stability analysis by shear strength reduction technique. *Soils and Foundations* 32(1) (1992), pp. 59-70.

[82] McKay, M. D., Conover, W. J., Beckman, R. J. A comparison of three methods for selecting values of input variables in the analysis of output from a computer code. *Technometrics* 21(2) (1979), pp. 239-245.

[83] Melchers, R. E. *Structural Reliability Analysis and Prediction*. 2nd edition. Chichester: John Wiley & Sons, 1999.

[84] Metropolis, N., Rosenbluth, A. W., Rosenbluth, M. N., Teller, A. H., Teller, E. Equations of state calculations by fast computing machines. *Journal of Chemical Physics* 21(6) (1953), pp. 1087-1092.

[85] Nataf, A. Determination des distribution dont le marges sont données, *Comptes Rendus de l'Académie des Sciences* 225 (1962), pp. 42-43.

[86] Nie, J., Ellingwood, B. R. Directional methods for structural reliability analysis. *Structural Safety* 22 (2000), pp. 233-249.

[87] Nie, J., Ellingwood, B. R. Finite element-based structural reliability assessment using efficient directional simulation. *Journal of Engineering Mechanics, ASCE* 131(3) (2005), pp. 259-267.

[88] Niederreiter, H. Point sets and sequences with small discrepancy. *Monatshefte für Mathematik* 104 (1987), pp. 273-337.

[89] Niederreiter, H. *Random Number Generator and Quasi-Monte Carlo Methods*. Volume 63 of *SIAM CBMS-NSF Regional Conference Series in Applied Mathematics*. Philadelphia: SIAM, 1992.

[90] Olsson, A., Sandberg, G., Dahlblom, O. On Latin hypercube sampling for structural reliability analysis. *Structural Safety* 25 (2003), pp. 47-68.

[91] Pandey, M. D. An effective approximation to evaluate multinormal integrals. *Structural Safety* 20 (1998), pp. 51-67.

[92] Papadrakakis, M., Papadopoulos, V., Lagaros, N. D. Structural reliability analysis of elastic-plastic structures using neural networks and Monte Carlo simulation. *Computer Methods in Appied Mechanics and Engineering* 136 (1996), pp. 145-163.

[93] Papaioannou, I., Der Kiureghian, A. Reliability-based design of slope angle considering spatial variability of soil material. In: *Proc. 6th International Conference on Computational Stochastic Mechanics*, Rhodos, June, 2010. Singapore: Research Publishing Services.

[94] Papaioannou, I., Heidkamp, H., Düster, A., Kollmannsberger, S., Rank, E., Katz, C. The subset simulation applied to the reliability analysis of a nonlinear geotechnical finite element model. In: *Proc. 7th International Probabilistic Workshop*, Delft, November, 2009. Delft: TU Delft.

[95] Papaioannou, I., Heidkamp, H., Düster, A., Rank, E., Katz, C. Towards efficient reliability methods with applications to industrial problems. In: *Proc. Ninth International Conference on Computational Structures Technology*, Athens, September, 2008. Stirlingshire: Civil-Comp Press.

[96] Papaioannou, I., Heidkamp, H., Düster, A., Rank, E., Katz, C. Integration of reliability methods into a commercial finite element software package. In: *Proc. 10th International Conference on*

Bibliography

Structural Safety and Reliability ICOSSAR 2009, Osaka, September, 2009. London: Taylor & Francis Group.

[97] Papaioannou, I., Heidkamp, H., Düster, A., Rank, E., Katz, C. Random field reliability analysis as a means for risk assessment in tunnelling. In: *Proc. 2nd International Conference on Computational Methods in Tunnelling EURO:TUN 2009*, Bochum, September, 2009. Freiburg: Aedificatio Publishers.

[98] Papaioannou, I., Straub, D. Geotechnical reliability updating using stochastic FEM. In: *Proc. 15th IFIP WG7.5 Working Conference on Reliability and Optimization of Structural Systems*, Munich, April, 2010. London: CRC Press, 2010.

[99] Papaioannou, I., Straub, D. Reliability updating in geotechnical engineering including spatial variability of soil. *Computers and Geotechnics* 42 (2012), pp. 44-51.

[100] Papoulis, A., Unnikrishna, S. *Probability, Random Variables and Stochastic Processes*. 4th edition. New York: McGraw-Hill, 2002.

[101] Phoon, K. K., Huang, H. W., Quek, S. T. Simulation of strongly non-Gaussian processes using Karhunen-Loève expansion. *Probabilistic Engineering Mechanics* 20 (2005), pp. 188-198.

[102] Pradlwarter, H. J., Schuëller, G. I., Koutsourelakis, P. S., Charmpis, D. C. Application of line sampling simulation method to reliability benchmark problems. *Structural Safety* 29 (2007), pp. 208-221.

[103] Proppe, C. Estimation of failure probabilities by local approximation of the limit state function. *Structural Safety* 30 (2008), pp. 277-290.

[104] Puig, B., Poirion, F., Soize, C. Non-Gaussian simulation using Hermite polynomial expansion: convergences and algorithms. *Probabilistic Engineering Mechanics* 17 (2002), pp. 253-264.

[105] Rackwitz, R. Reviewing probabilistic soils modelling. *Computers and Geotechnics* 26 (2000), pp. 199-223.

[106] Rackwitz, R. Reliablity analysis - a review and some perspectives. *Structural Safety* 23 (2001), pp. 365-395.

[107] Rackwitz, R., Fiessler, B. Structural reliability under combined random load sequences. *Computers and Structures* 9(5) (1978), pp. 484-494.

[108] Rajashekhar, M. R., Ellingwood, B. R. A new look at the response surface approach for reliability analysis. *Structural Safety* 12 (1993), pp. 205-220.

[109] Rocco, C. M., Moreno, J. A. Fast Monte Carlo reliability evaluation using support vector machine. *Reliability Engineering and System Safety* 76 (2002), pp. 237-243.

[110] Rosen, J. B. The gradient projection method for nonlinear programming. Part II. Nonlinear constraints. *Journal of the Society for Industrial and Applied Mathematics* 9 (1961), pp. 514-532.

[111] Rosenblatt, M. Remarks on a multivariate transformation. *Annals of Mathematical Statistics* 23 (1952), pp. 470-472.

[112] Rubinstein, R. Y. *Simulation and the Monte Carlo Method*. New York: John Wiley & Sons, 1981.

[113] Saff, E. B., Kuijlaars, A. B. J. Distributing many points on a sphere. *Mathematical Intelligencer* 19 (1997), pp. 5-11.

[114] Sakamoto, S., Ghanem, R. Polynomial chaos decomposition for the simulation of non-Gaussian nonstationary stochastic processes. *Journal of Engineering Mechanics, ASCE* 128(2) (2002), pp. 190-201.

[115] Santosh, T. V., Saraf, R. K., Ghosh, A. K., Kushwaha, H. S. Optimum step length selection rule in modified HL-RF method for structural reliability. *International Journal of Pressure Vessels and Piping* 83 (2006), pp. 742-748.

Bibliography

[116] Schall, G., Gollwitzer, S., Rackwitz, R. Integration of multinormal densities on surfaces. In: *Proc. 2nd IFIP WG7.5 Working Conference on Reliability and Optimization of Structural Systems*, London, September, 1988. New York: Springer-Verlag, 1988.

[117] Schittkowski, K. On the convergence of a sequential quadratic programming method with an augmented Lagrangian line search function. *Mathematische Operationsforschung und Statistik, Series Optimization* 14 (1983), pp. 197-216.

[118] Schuëller, G. I., Pradlwarter, H. J., Koutsourelakis, P. S. A critical appraisal of reliability estimation procedures for high dimensions. *Probabilistic Engineering Mechanics* 19(4) (2004), pp. 463-474.

[119] Schuëller, G. I., Stix, R. A critical appraisal of methods to determine failure probabilities. *Structural Safety* 4 (1987), pp. 293-309.

[120] Schueremans, L., Van Gemert, D. Benefit of splines and neural networks in simulation based structural reliability analysis. *Structural Safety* 27 (2005), pp. 246-261.

[121] Schweiger, H. F., Peschl, G. M. Reliability analysis in geotechnics with the random set finite element method. *Computers and Geotechnics* 32 (2005), pp. 422-435.

[122] Shinozuka, M., Deodatis, G. Simulation of stochastic processes by spectral representation. *Applied Mechanics Reviews* 44(4) (1991), pp. 191-204.

[123] Shinozuka, M., Deodatis, G. Simulation of multi-dimensional Gaussian stochastic fields by spectral representation. *Applied Mechanics Reviews, ASME* 49(1) (1996), pp. 29-53.

[124] Smith, I. M., Griffiths, D. V. *Programming the Finite Element Method*. 4th edition. Chichester: John Wiley & Sons, 2004.

[125] Sobol, I. M. The distribution of points in a cube and the approximate evaluation of integrals. In Russian. *USSR Computational Mathematics and Mathematical Physics* 7 (1967), pp. 784-802.

[126] SOFiSTiK AG. *SOFiSTiK analysis programs version 2010.* Oberschleißheim: SOFiSTiK AG, 2010.

[127] Spanos, P. D., Ghanem, R. G. Stochastic finite element expansion for random media. *Journal of Engineering Mechanics, ASCE* 115(5) (1989), pp. 1035-1053.

[128] Stefanou, G., Papadrakakis, M. Assessment of spectral representation and Karhunen-Loève expansion methods for the simulation of Gaussian stochastic fields. *Computer Methods in Appied Mechanics and Engineering* 196 (2007), pp. 2465-2477.

[129] Stewart, M. G., Melchers, R. E. *Probabilistic Risk Assessment of Engineering Systems.* London: Chapman & Hall, 1997.

[130] Straub, D. Reliability updating with equality information. *Probabilistic Engineering Mechanics* 26(2) (2011), pp. 254-258.

[131] Straub, D., Lentz, A., Papaioannou, I., Rackwitz, R. Life quality index for assessing risk acceptance in geotechnical engineering. In: *Proc. 3rd International Symposium on Geotechnical Safety and Risk,* Munich, June, 2011. Karlsruhe: Bundesanstalt für Wasserbau.

[132] Sudret, B., Der Kiureghian, A. *Stochastic Finite Elements and Reliability: A State-of-the-Art Report.* University of California, Berkeley, 2000 – Technical Report no UCB/SEMM-2000/08.

[133] Tang, L. K., Melchers, R. E. Improved approximation for multinormal integral. *Structural Safety* 4 (1987), pp. 81-93.

[134] Todor, R. A. Robust eigenvalue computations for smoothing operators. *SIAM Journal on Numerical Analysis* 44(2) (2006), pp. 865-878.

Bibliography

[135] Tschebotarioff, G. P. *Soil Mechanics, Foundations and Earth Structures*. New York: McGraw-Hill, 1951.

[136] Tvedt, L. Distribution of quadratic forms in normal space - application to structural reliability. *Journal of Engineering Mechanics, ASCE* 116(6) (1990), pp. 1183-1197.

[137] Vanmarcke, E. H. *Random Fields: Analysis and Synthesis*. Cambridge, MA: MIT Press, 1983.

[138] Vanmarcke, E. H., Grigoriu, M. Stochastic finite element analysis of simple beams. *Journal of Engineering Mechanics, ASCE* 109(5) (1983), pp. 1203-1214.

[139] Von Neumann, J. Various techniques used in connection with random digits. *Applied Mathematics Series* 12 (1951), pp. 36-38.

[140] Yamazaki, F., Shinozuka, M. Digital generation of non-Gaussian stochastic fields. *Journal of Engineering Mechanics, ASCE* 114 (1988), pp. 1183-1197.

[141] Zienkiewicz, O. C., Cormeau, I. C. Visco-plasticity – plasticity and creep in elastic solids – A unified numerical solution approach. *International Journal on Numerical Methods in Engineering* 8(4) (1974), pp. 821-845.

[142] Zhang, Y, Der Kiureghian, A. Dynamic response sensitivities of inelastic structures. *Computer Methods in Appied Mechanics and Engineering* 108 (1993), pp. 23-36.

[143] Zhang, Y, Der Kiureghian, A. Two improved algorithms for reliability analysis. In: *Proc. 6th IFIP WG7.5 Working Conference on Reliability and Optimization of Structural Systems*, Assisi, September, 1994. London: Chapman & Hall, 1995.

[144] Zhang, J., Ellingwood, B. Orthogonal series expansions of random fields in reliability analysis. *Journal of Engineering Mechanics, ASCE* 120(12) (1994), pp. 2660-2677.

Bibliography

i want morebooks!

Buy your books fast and straightforward online - at one of world's fastest growing online book stores! Environmentally sound due to Print-on-Demand technologies.

Buy your books online at
www.get-morebooks.com

Kaufen Sie Ihre Bücher schnell und unkompliziert online – auf einer der am schnellsten wachsenden Buchhandelsplattformen weltweit! Dank Print-On-Demand umwelt- und ressourcenschonend produziert.

Bücher schneller online kaufen
www.morebooks.de

VDM Verlagsservicegesellschaft mbH
Heinrich-Böcking-Str. 6-8   Telefon: +49 681 3720 174   info@vdm-vsg.de
D - 66121 Saarbrücken   Telefax: +49 681 3720 1749   www.vdm-vsg.de

Printed by Books on Demand GmbH, Norderstedt / Germany